SOUND HEALING & VALUES VISUALIZATION

Creating a Life of Value

SOUND HEALING & VALUES VISUALIZATION

Creating a Life of Value

JOHN BEAULIEU, N.D., PH.D.

Copyright © 2018 by John Beaulieu. All rights reserved.

Published by BioSonic Enterprises, Ltd.
High Falls, New York 12440

Design and production: Pamela Kersage
Cover design: Philippe Garnier
Editing: Thea Keats Beaulieu
Photographs: Lars Beaulieu
Photo of John Beaulieu in concert in Zurich, Switzerland: Christine Pieler
Photo of John Beaulieu in concert in Zug, Switzerland: Jorgos Ledermann

"The Perfect Fifth—The Science and Alchemy of Sound" is published online in the *Rose+Croix Journal* Vol. 11, www.rosecroixjournal.org, and is reproduced with permission of the *Rose+Croix Journal*.

Cymatics photographs are courtesy of MACROmedia

Manufactured in the United States of America

ISBN: 978-1-5323-7780-8

The lessons taught in *Sound Healing and Values Visualization* do not give you legal permission to diagnose, prescribe, treat, or engage in sound healing for any human illness or condition without qualifying in your local jurisdiction, and/or obtaining a license or certification, if necessary. Each state and county has different laws concerning the practice of healing and health providing. You should consult your local association or attorney before you provide any services.

Contents

Foreword..vii
Introduction...xi

SECTION ONE: Sound...1
    PART 1: Sacred Sound Science...3
    PART 2: Mindful Listening...13
    PART 3: Be Like a Child...21
    PART 4: Listen Like a Scientist.......................................27
    PART 5: Sound Stories to Warm Your Ears...............................35

SECTION TWO: Sound Healing and Values Visualization Process...............43
    PART 1: Visualization...45
    PART 2: Values..51
    PART 3: Feeling Tones...57
    PART 4: Sound and Values Visualization................................69

SECTION THREE: Sound Healing and Values Visualization Practice............91
    How to Use Sound to Work with Dependencies............................91
    Working with Cancer and Health Challenges.............................98
    Creating Your Perfect Living Space...................................100
    Finding Your Ideal Life Partner......................................104
    Winning: Visualizations for Success..................................107
    Guided Wellness Visualizations.......................................110

SECTION FOUR: Sound Musings..121
    The Perfect Fifth: The Science and Alchemy of Sound..................121
    Annotated Nada Bindu Upanishad.......................................146

SECTION FIVE: Appendices...149
    APPENDIX A: How to Sound BioSonic Tuning Forks.......................149
    APPENDIX B: Human Tune In™ "A BioSonic Sound Healing Concert"........151
    APPENDIX C: Anchoring and Values Visualization.......................155
    APPENDIX D: Five Element Hand Shaking Exercise.......................159
    APPENDIX E: Element Advertisement Examples...........................161
    APPENDIX F: Hypnosis and Element Integration Examples................162

Endnotes...163
Bibliography...171
Acknowledgements...177

Dreaming the Light

Everybody has a dream of light.

Supporting another's light dream is to support your own light dream.

A dream of light spins dark material into newer and newer

illuminated patterns.

The dark used to be a time to hide, a time to fear,

a time to huddle together for warmth.

Light dreamers go into the dark and emerge with material to burn

in their inner Fire.

They understand that strands of darkness can be transformed

into dreams of light.

Foreword

In the period of time since the publication of *Human Tuning: Sound Healing with Tuning Forks* in 2010, the use of sound as a therapeutic agent has grown exponentially throughout the world. John Beaulieu, N.D., Ph.D., has been at the forefront of this movement for the past 45 years and has contributed significantly with the introduction of his BioSonic tuning forks and his several books and publications. *Human Tuning* has been an enormous contribution to sound healing practitioners because it operationalizes the idea that we are sonic beings that can get easily out of tune and, like musical instruments, must be tuned regularly for good performance. In *Human Tuning*, Dr. Beaulieu introduced a simple, portable tool (tuning forks) with a series of protocols that, regardless of musical knowledge or training, can be used by anyone in group, individual, or self-administered sessions as a tool to facilitate relaxation, meditation, and centering. The tuning forks are also tools that can be used to facilitate psycho and somatic therapies and bodywork, social communication, and productivity in social and work settings.

In *Sound Healing and Values Visualization*, Dr. Beaulieu emphasizes the role of consciousness/self-awareness in creating and cementing the therapeutic effects of sound healing practices. *The essential task of sound healing practices consists of "creating a healing intention for sound."* Intentionality is the defining characteristic that differentiates living from inanimate matter. All living organisms from the simplest to the most complex exhibit intentionality. In the past, intention has always been spoken of in healing practices as the essential quality or attitude ("for the greater good") that healers must possess if they are to be successful. In his current work, Dr. Beaulieu turns the focus back to where it most belongs, the person receiving the treatment. *Sound Healing and Values Visualization* introduces a method for individuals to clarify, fine tune, and visualize who they are and what they want to achieve in their lives, and it presents a simple-to-use set of practices using sound, mindful listening, and meditative techniques. This allows for a transformation and redirection of consciousness and nervous activity which is conducive to wellness and optimal well being.

Given the rapid expansion of sound healing events in the past decade and the seeming lack of direction, clinical understanding, and experienced practitioners working with individuals using sound therapeutically, it cannot be understated how timely this current work is. Dr. Beaulieu demonstrates the process of Values Clarification and Visualization followed by the use of sound and affirmations using simple-to-use Mindful Listening and meditative techniques. These processes create a truly therapeutic experience for the individual and not just a fleeting feel-good moment. There is now a significant body of research on mindfulness meditation illustrating how simply quieting the mind regularly can produce physical, physiological, emotional, behavioral, and socio-behavioral changes in individuals.[1] Quieting the mind, however, can be difficult when it is plagued by distortions that prevent a clear vision and understanding of the individual's own sense of self and goals. *Values Clarification addresses these distortions and, used along with mindful listening and meditative techniques, sets the stage for the visualization of the healing intention in sound that can transform consciousness in a therapeutic manner.*

The theoretical foundations supporting the use of sound in a therapeutic manner stem from the fact that sound and movement are an integral part of life and the physical world and have influenced the course of all inanimate and animate (biological/social) processes and evolution.[2] Human beings are sonic entities designed to produce and process sounds, vibration, and movement. We are multi- and inter-dimensional beings that live in different dimensions and planes of existence simultaneously: in the cochlea, sound is perceived in both physical and subatomic (quantum) space. *The perception of sound is, therefore, both a physical and an ethereal or spiritual event and consciousness is the instrument we use to navigate and balance movement between these planes.*

All living organisms have evolved in relationship with and under the influence of sound, movement, and vibration. The earliest and simplest living organisms developed structures to sense their environment and to move towards or away from entities, places, events, or *known* threats. These organisms possessed a sensorium, a motorium, and the ability to process and store information (memory) despite the fact that *they had no brains or nervous systems.*[3,4,5] The earliest biological life forms, therefore, demonstrated *intentionality* and *learning capabilities* that cannot be explained convincingly without postulating the existence of a consciousness behind the behavior.

The existence of consciousness is what differentiates animate from inanimate matter, the living from the non-living. Unlike inanimate matter, living organisms act upon the environment with intention that creates changes in the quantum field of subatomic particle waves. The will to survive, to maintain balance internally and with the external environment, to self-repair, is observed only in living organisms. This innate and intrinsic drive to establish and maintain homeostasis and survive is the essence and definition of healing. Healing is therefore a defining characteristic of life and a gateway into an understanding of the nature of consciousness as manifested by the ability of simple organisms without brains, nervous systems, or organized genetic material to survive by making choices (intentionality) and learning from previous experience. Becoming aware of, focusing, and manifesting intention(s) is a primary task of all living beings and Dr. Beaulieu taps into that primal process and shows us a way to facilitate it using sound and vibration with his BioSonic tuning forks and mindful techniques. I highly recommend *Sound Healing and Values Visualization* to anyone interested in finding out how to use sound in this way.

— David Perez-Martinez, M.D.

Introduction

In *Sound Healing and Values Visualization*, I present an integrative sound healing process that I have developed and worked with for over forty years. The process is an extension of my book *Human Tuning: Sound Healing with Tuning Forks* which was designed to work with and enhance all sound healing practices. In *Human Tuning*, I wrote:

> Visualization is the act of creating an intention for the sound. The visualization to heal has to be clear before the forks are tapped. Method without visualization is limited. Through visualization, the sound of the tuning forks becomes a deeper healing experience.[1]

In this new book, *Sound Healing and Values Visualization*, I present a systematic and practical understanding of intention, visualization, and sound based on value. The Sound Healing and Values Visualization Process integrates visualization, values clarification, and feeling tone psychology with sound and music. It is designed to create a highly tuned mental event focused on healing and the attainment of life goals. When a focused intention is carried by sound into the core brain, the brain activates multi-dimensional neural networks and searches for new formations that lead to healing and the obtainment of your life's goals. The Values Visualization Process can be used to set and reach your goals. It can be used with your family, friends, students, and clients to enhance healing and goal achievement.

Sound Healing and Values Visualization is divided into five sections. Each section can be read individually, but they are also each part of the whole.

1. The first section, "Sound," explores how sound mimics the vibratory nature of existence that sustains and imbues everything that exists within structure and form. It then shows practical ways of exploring the vibratory nature of existence through mindful listening, creative sound explorations, and research based on the inner experience of sound explorations.

The physicians of the past worked from a vibrational model of the universe similar to our modern understanding of a vibrational quantum field. They understood that everything was interconnected. Their professional language was one of archetypes and the application of those archetypes to the challenges of everyday life. Comparing their archetypical language to today's modern stress science reveals the depth of their understanding of stress and their ability to work with stress-related diseases. Hans Selye said in his book *The Stress of Life* that his general adaptation syndrome could have been discovered in the Middle Ages or earlier through an unbiased state of mind.[2] Although these doctors did not have modern biochemistry or scientific procedures, they were nevertheless doctors who closely observed their patients and learned from their behaviors. We have much to learn from the physicians and healers that have come before us. They left us with a puzzle that needs to be pieced together by our modern scientific understandings. We may never know exactly how those systems were practiced, but we can learn from our experiences inspired by their work and apply them to evidence-based practices. Ultimately, the clinical outcome, whether it be through phenomenological experience, reductionist science, or systems integration, will always suggest the need for more research.

2. The second section, "Sound Healing and Values Visualization Process," presents the process and how it can be used to create highly focused and individualized healing outcomes that integrate with proven evidence-based practices. Psychological research suggests that visualizing and working towards your dreams and goals is related to positive emotions and overall enhanced well being. Dr. Bettina Wiese of University of Zurich summarized the current research in these words, "…empirical research has repeatedly shown that striving toward self-concordant goals strengthens the link between goal progress and well-being."[3] Self Determination theory suggests that the more we internalize and identify with a goal (dream) by understanding the value behind it, the more likely we are to be motivated to act.[4] When we do not understand our goals, dreams, and values, our life can lose meaning and become like a daily task dictated from outside of us. When this happens we procrastinate and put things off, which can lead to a lack of motivation and self worth, which can lead to depression. Understanding and making progress towards our dreams and goals leads us to more positive emotions and greater satisfaction with our lives.

When our well-being increases, we become internally motivated to act on and manifest our dreams.

3. The third section, "Sound Healing and Values Visualization Practice," presents specific step-by-step examples of the Sound Healing and Values Visualization Process. You will find protocols and guided meditations for different conditions that you can immediately use.

4. The fourth section, "Sound Musings," includes two articles: *The Perfect Fifth: The Alchemy of Sound,* an article I wrote for the *Rose+Croix Journal* on the Perfect Fifth,[5] and my annotations of the *Nada Bindu Upanishad.* I have included these articles and my Upanishad annotations to honor the sound healing and vibrational insights of those who came before me. I have also included these writings because Sound Healing and Values Visualization was inspired by my clinical work with tuning forks. This is discussed in depth in *Human Tuning: Sound Healing with Tuning Forks.*

Tuning forks are unique in the field of sound healing. Everything in this book can be successfully practiced with BioSonic tuning forks. Body Tuners™ tuned to the Perfect Fifth or to C or G are versatile and neutral sound healing instruments. They are versatile because they are always in tune, are lightweight, can be easily carried in your pocket, and are easy to learn and use. They are neutral because they are used by musicians all over the world to tune their instruments and are not associated with any specific culture or style of music. Tuning forks lend themselves to research and have consistent results with clients because of their tuning accuracy. In general, when tuning forks are tapped with healing intentions, their effects are quick and can be integrated with and enhance many forms of therapy. Hospitals are stressful; doctors, nurses, therapists, and support staff have a lot to do. Tapping C & G tuning forks, just for a moment, is a way for those practitioners to shift gears with a patient and then move on to the next patient, knowing that the sound will serve to enhance whatever therapies the patient is receiving.

The primary tuning forks used in the world today by sound healers are BioSonic tuning forks. BioSonic tuning forks include vibroacoustic (Otto) and psychoacoustic forks. The primary BioSonic tuning forks are the C256 cps, the G384 cps, and the Otto 128 cps. As part of the Fibonacci series, the C & G create an interval of a Perfect Fifth (a perfect 2/3 ratio) when sounded. The Otto 128 cps (the difference tone obtained by subtracting C256 from G384) spreads the vibration of the

Perfect Fifth vibroacoustically throughout the body when placed directly on any tissue and is especially effective when pressed directly over a bone. The goal of treatment is to create a state of balance by entraining the person's nervous system to the vibratory rate of a Perfect Fifth. The sound (psychoacoustic) and vibratory (vibroacoustic) effects of this entrainment operate by triggering a relaxation response that is hypothesized to be related to the nitric oxide stimulation observed with the use of these tuning forks.[6]

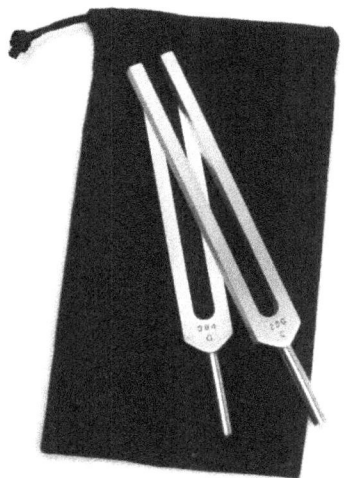
BioSonic BodyTuners™

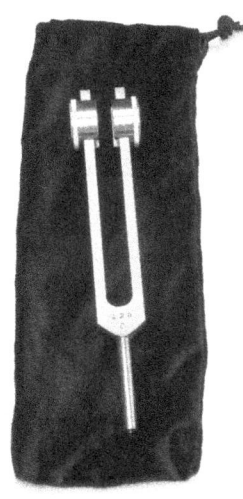
BioSonic Otto 128™

Dr. Beaulieu practicing values visualization with C & G Body Tuners™

Appendix A gives specific instructions on how to use BioSonic tuning forks. If you want to know more about how I discovered tuning forks as sound healing instruments, as well as research on the positive effect of sounding tuning forks, you can find it in *Human Tuning: Sound Healing with Tuning Forks.*

John Beaulieu, N.D., Ph.D.
Stone Ridge, New York
2018

SECTION ONE
Sound

If you wish to understand the universe, think of energy, frequency and vibration.
— Nikola Tesla

Understanding sound and working with different sounds is fundamental to the practice of sound healing. When we immerse ourselves in sound, we go beyond normal rational meanings, beliefs, and psychological defenses into a vibratory universe of flowing sensory experience. Our ability to conduct sensory vibrations through our mind and body will ultimately determine our state of mind. The more vibrations we can conduct, the more we will be able to adapt to continually changing conditions. There is an infinite number of vibrations and combinations of vibrations that are always available to us. It is important for us to develop our abilities to explore and creatively adapt to different vibratory states.

This section is presented in five parts. Each part can be understood separately. However, in sound healing practice they merge, overlap, and integrate in many creative ways and orders.

PART 1: Sacred Sound Science

PART 2: Mindful Listening

PART 3: Be Like a Child

PART 4: Listen Like a Scientist

PART 5: Sound Stories to Warm Your Ears

Prelude

We live in an oscillating universe of infinite possibilities.

We are vibrational beings. We resonate.

We travel through multi-dimensional entangled filaments of light.

We are navigators on a vast ocean of fluidic ambrosia.

We hyper-leap through synaptic vortexes

surrounded by entangled webs of illuminated light filaments.

We are light beings of infinite possibilities dancing

in a cosmic womb of stillness to the sounds of a musical universe.

PART 1
Sacred Sound Science

Sound is the force of creation, the true whole. Music then, becomes the voice of the great cosmic oneness and therefore the optimal way to reach this final state of healing.

— Sufi Inayat Khan

Sound has been used continuously throughout time and in all cultures as a therapeutic agent in medical and healing practices. Healing with sound and music is recognized in all of the traditional natural healing systems, including Oriental, Ayurvedic, Tibetan, and Hermetic Medicine. The ancient Rishis of India created mantras as a means of expanding consciousness and obtaining high-level wellness, which they called enlightenment. Based on Pythagorean mathematics, the ancient Greeks used sonic intervals to positively affect consciousness in order to stimulate dreams and heal traumas. Shamans from all indigenous traditions have used vocal sounds, whistles, rattles, drums, didgeridoos, flutes, and other kinds of sound making instruments in order to enter into altered states of consciousness in their quest to heal individuals and entire communities. These traditions have in common a universal view of consciousness and healing based on an understanding of sound and vibration.

Sound healing is a special field of study and practice that integrates scientific disciplines with one's intuition and creative inspiration. Writing in *Nutrition and Integrative Medicine: A Primer for Clinicians,* John Beaulieu, N.D., Ph.D., and David Perez-Martinez, M.D., defined sound healing as the practice of using sound and listening in a mindful manner to transform and expand consciousness in order to enhance our body's natural drive to regenerate and heal itself. The basic premise and theoretical foundation of sound healing is that *All existence is vibratory in nature* and, therefore, *It is the underlying vibratory field that sustains and imbues everything that exists with structure and form.* Sound is the perfect tool for healing practices and personal growth because it *mimics the vibratory nature of existence* and affects individuals on all levels: anatomical, physiological, emotional, psychological, and spiritual.[1]

Physicists call the universal vibrational field that sound mimics the *quantum field*. The quantum field is a continuous vibrational medium which is present everywhere in space. It is an unbroken interconnected whole that can be imagined as a great ocean of waves and currents of unlimited creative potential. The physicist Werner Heisenberg said, "The atoms or elementary particles themselves are not real; they form a world of potentialities or possibilities rather than one of things or facts."[2]

The discovery of the quantum field is based on an understanding of sound. In order to better understand atoms, physicists visualized them as musical instruments such as organs, pianos, and bells. *Like a note on a piano each atom was tuned to its own unique frequency, creating a keyboard of elements.* This visualization created new questions such as "What tunes each atom?" and "How and why do electrons jump from one orbital path to another, rather then sliding from one path to another similar to a glissando in music?" Max Planck, a German physicist and accomplished pianist, used the music and sound visualization of an atom to answer these questions. He imagined that electron orbits were vibrating strings that behaved like standing waves.

A standing wave is a wave that happens between fixed points. Imagine holding a rope attached to a tree. The tree is fixed and you are fixed. When you flick the rope it creates a wave that travels to the tree. This wave is called a traveling wave because if the tree were not there it would theoretically continue and never return. However, when the wave reaches the tree, it returns back to you. This creates a specific frequency standing wave as long as you keep flicking the rope. The slower you flick the rope the lower the frequency and the faster you flick the rope the higher the frequency.

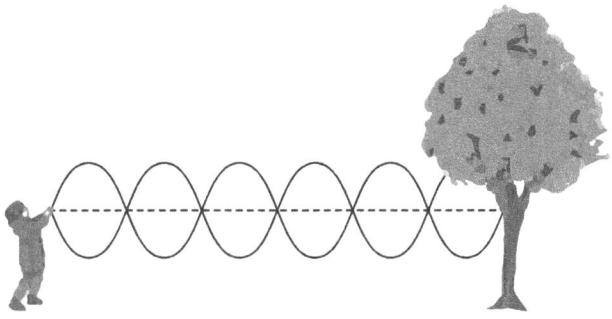

When a string vibrates, all of its energy is contained between its nodes. Listening to a sound, beyond our labels, associations, and beliefs about the sound, is a mental event that is defined by frequency and nodes. A string cannot vibrate without nodes and nodes cannot exist without a vibrating string. The two can be talked about separately, but they are one and the same.

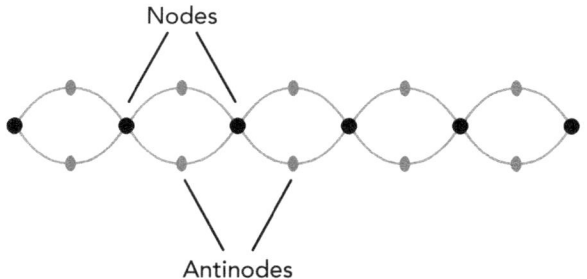

In the above diagram and in most all two-dimensional diagrams of sound waves, nodes are represented by black dots or lines crossing. This is visually misleading because it gives you the impression that a node is a dot on the paper. Nodes are not dots. Nodes are no-thing. In quantum chemistry, a node is a place where all points have no density. Nodes have no meaning and yet without them all realities would cease to exist and all vibrational beings in those realities would cease to exist. Nodes are holes that are necessary to the formation of continually changing harmonic networks. In a two-dimensional representation, nodes may appear to change. They never actually change, though, because it is impossible for "nothing to change."

Antinodes are the measurable frequency vibration of a wave, which is the polar opposite of a node. Antinodes are "something" and are continually changing, based on different frequencies. When we change frequencies, the length of the wave can become longer or shorter causing the two-dimensional "node dots" to appear further apart or closer together. This is an illusion. If our awareness is located inside a node, nothing will change; however, if our awareness is located inside the frequency of the antinode, everything will change. This mirrors the relationship between sound and silence. Sound is always changing and silence can appear and disappear at different times and places but it is always silent.

The composer Claude Debussy said that music is the silence between the notes. When I took a master piano class with Sophia Rosoff, she emphasized the importance of the silent spaces around the sounds. For her, a music composition had very little to do with notes. Playing more notes or even playing the right notes was secondary to the awareness of the infinite noteless space around and between the notes. To my surprise, we worked very hard on silence. Notated rests were merely the infinity of soundless space making itself known through the sound of the music. Notes were just ways of directing the mind to tune into and listen to infinite silence.

The following pictures are from a video taken at Argonne Labs.[3] There are two speakers, one at the bottom and another at the top, that produce an infrasound standing wave that can be seen on an oscilloscope. An oscilloscope is a device for viewing oscillations on a screen. The scientist uses a dropper to place drops of water in the nodes. The drops of water are suspended in space by the power of the infrasound wave.

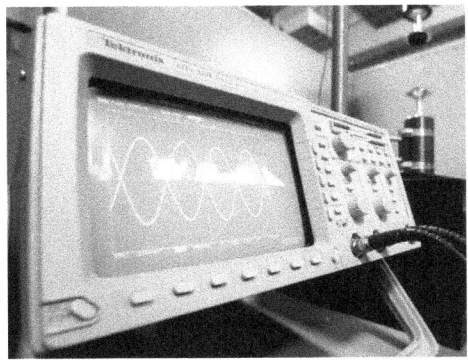

Infrasound Waves displayed on oscilloscope.
The water droplets are places where the wave cross.

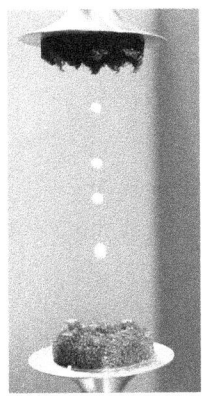
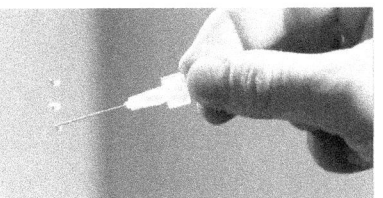

Max Planck knew that when a string vibrates, all of its energy is quantized between the nodes in precise units. He found that these units corresponded exactly to the electromagnetic frequencies emitted during the transition of electrons between energy levels within an atom. Each atom has a unique spectrum of energy between its inner proton core and electron orbits, much like chords in music consist of unique tones. He discovered that the numbers corresponded exactly with the atomic spectra model of Niels Bohr. This was the origin of Planck's quantum theory for which he was awarded the Nobel Prize in 1918.

Louis de Broglie, a physicist and accomplished violinist, visualized atoms as stringed instruments. He reasoned that if the circular strings of atoms vibrated, then they must produce harmonics like the strings of a violin. Planck visualized electrons as spinning particles creating a standing wave frequency and de Broglie visualized electrons as tones with harmonics. This meant that if electrons behaved like strings, they were both waves and particles. At the time, de Broglie's harmonic theory of atoms was dismissed and considered absurd. Not long after, it was shown that irradiating atomic nuclei with energy produces harmonics similar to those of a vibrating string. In 1929, de Broglie was awarded the Nobel Prize.

What Planck and de Broglie failed to understand was that we are vibrational beings living in a vibrational universe of waves and nodes. Before the term "quantum" was invented, the sages of the East and West set forth models of a vibrational universe. They used sounds we hear with our ears to enter into the larger experience of the vibrational dynamics of the whole universe. They understood that, through direct experience, sound mimics the vibrational nature of reality. By mindfully listening to sounds, they learned about our innermost vibrational relationship with the universal field of vibration. This learning involved living in resonance with the vibrational nature of reality.

The harmonics explored by de Broglie at the quantum level are called *overtones* in music. Overtones are tones that appear in a precise mathematical sequence after the sounding of a fundamental tone. Overtones are the reason that musical instruments can have the same frequency and yet sound different. For example, the note C played on a flute, piano, violin, or trumpet is the same note, but it makes a different sound. This is because each instrument makes different harmonics louder or softer, which gives each instrument its own unique tone. Musicians call this tone color. All sound healing and traditional musical instruments, i.e., gongs, crystal bowls, bells, whistles, drums, flutes, pianos, our voice, etc., create different harmonics. Many people say that the harmonics of different instruments are like the personality or spirit of an instrument. For example, musicians and sound healers will spend thousands of dollars for instruments that resonate certain overtones that "fit their ear."

The overtone series ascends in whole numbers, i.e., 1 (fundamental), 2, 3, 4, 5, 6, 7, 8, 8, 9, …n. When we pluck a string, the string divides again and again to create many tones. The primary sound we hear is called the fundamental tone which is represented in the diagram as an undivided wave. The fundamental tone continues to divide, theoretically creating an infinite number of overtones.

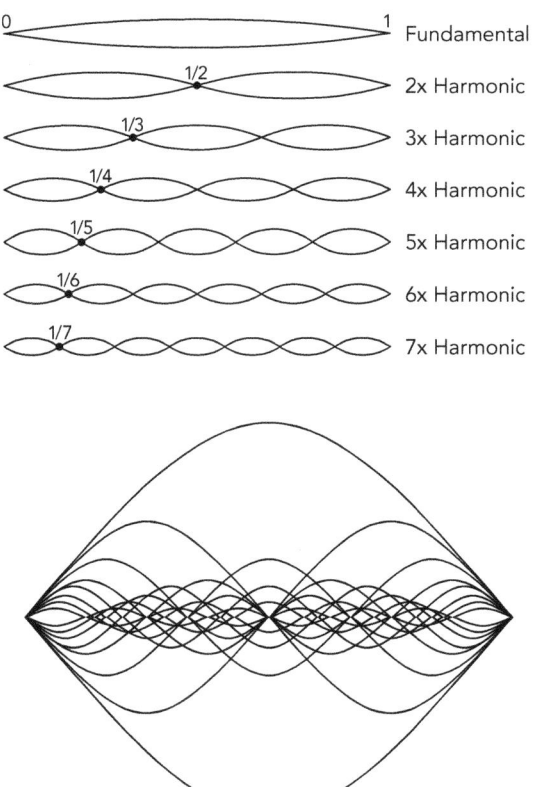

The overtone series creates waves within waves of sound.

In 2013, Clemson University researchers released a video project called *Shape Oscillation of a Levitated Drop in an Acoustic Field*.[4] In this video, an ultrasonic standing wave field was created where water was inserted at the nodes. Next, the resonant frequency of the water drops levitated in the nodes and was matched to the strength of the acoustic field. Using field frequency modulation, they were able to show that the behaviors of the water droplets within the nodes form different harmonics. They discovered that the water droplets formed pulsating star patterns and that the number of points on the star matched the field harmonic.

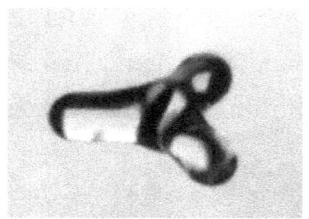
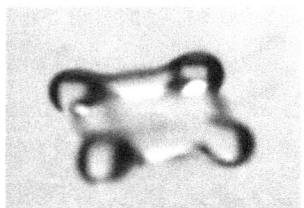

3rd Harmonic   4th Harmonic

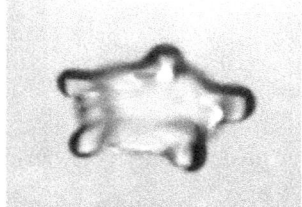

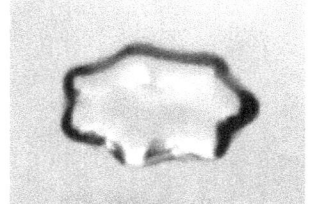

5th Harmonic   6th Harmonic   7th Harmonic

Our body is composed of "protein harmonics." Proteins are the building blocks of tissues and organs, and they also regulate how genes are expressed. One of the greatest challenges in biochemistry is to learn how proteins fold to become biologically functional three-dimensional structures. In quantum physics, we also see the emergence of a biochemical harmonic understanding of matter in order to solve the problem. The technical name for this is *protein sonification*. Researchers transform data about proteins into musical sounds. The discoverer of this phenomena, Dr. Christian B. Anfinsen, winner of the Nobel Prize in biochemistry, said of his discovery, "It struck me recently, that one should really consider the sequence of a protein molecule about to fold into a precise geometric form, as a line of melody written in a canon form & so designed by Nature to fold back into itself, creating harmonic chords of interaction consistent with biological function."[5] Anfinsen is telling us that geometric form and harmonics are one and the same. If you change the "harmonic chords," you will change the biological function and then consequently the geometric shape. In the Nov. 18, 2016, edition of *Science* one can listen to the music of proteins folding.[6]

Playing, singing, and listening to overtones has long been a method of healing and expanding conscious awareness. For centuries, Buddhist chanters in Mongolia and Tibet have been singing overtones embedded with sacred mantras. They resonate harmonics in different sinus cavities and spaces within their craniums while chanting. Many cultures have produced singing sounding bowls and gongs. When these are tapped or rubbed they start to produce different harmonics. Today, the most well known sounding bowls are Tibetan Singing Bowls and Crystal Bowls. In the early 1900's, the Russian composer Alexander Scriabin believed that sounding specific overtones, which he called mystic chords, would bring forth a new era and unite Heaven and Earth.

Scriabin's *Mystic Chord*

Scriabin's last composition, *Mysterium*, was to be played in India using etheric bells hung from the clouds that sounded very high-pitched overtones, like wind chimes. Scriabin visualized himself sitting on earth listening to harmonic voices of angelic beings called Devas, and these Devas would bring forth a new era of enlightenment. The intervallic spaces between overtones were known by the ancient Taoists as the Mysterious Mountain Passageways leading to angelic kingdoms. William Blake's *Jacob's Ladder* (1800) is a continual spiral, ascending from Earth into higher and higher realms.

Jacob's Ladder, 1805, British Museum, London

The image of Jacob's Ladder is often associated with the ascension of overtones. Jacob's Ladder begins on earth and rises to heaven. Earth is a metaphor for the fundamental tone and each step of the ladder represents a different overtone ascending to heaven. In 1992, I met with the Kabbalah scholar Rabbi Berg. During our conversation, I mentioned the mystical concept that harmonics were a stairway to heaven. His eyes lit up and he said that in the Kabbalah there are names for angels that live in the intervals created by each harmonic. Rabbi Berg did not say more and left with me with that mystic mystery.

When I was a teenager in 1963, I lived in Atlanta, Georgia, for a summer with my uncle, who taught me how to play the harmonica. I learned songs like "Red River Valley," "Clementine," "Aura Lee," and "She'll Be Coming 'Round the Mountain." These were songs that were composed using traditional Western scales and harmonics. My uncle was a professional contractor and many of his crew were black laborers. The South was segregated at the time and fortunately my uncle was against segregation and socialized with his crew. One night, he took me to visit one of the workers who lived in a row house outside of Atlanta. It was a country setting and there were musicians playing on the back porch. They played blues and it was the first time in my life I had heard that kind of music. When they started bending notes on their harmonicas and guitars, chills ran up my spine and over my whole body. I didn't understand at the time that they were playing overtones that were not accepted in our normal Western music. I thought the top of my head was going to lift off. I went home and spent hours bending notes on my harmonica. Today, people of all ages can play an electric keyboard and bend sounds into different overtones with a simple pitch wheel.

For Planck and de Broglie, sound was the analogy they used to solve a problem. They were not prepared to embody the vibrational nature of the quantum field through a direct experience of sound. It was not surprising that the quantum behaviors physicists discovered at the subatomic level could not be reconciled with what they understood as logical reality. The physicist Niels Bohr said, "Anyone who is not shocked by quantum theory has not understood it."[7] The "shocking" quantum behaviors that Niels Bohr is referring to are called non-locality, uncertainty, and wave-particle duality. Non-locality, which is also called quantum entanglement, occurs when quantum particles remain connected at vast distances so that actions performed on one can immediately affect the other. Albert Einstein called this phenomena of non-locality "spooky action at a distance." Uncertainty is when the observer has an effect on what is being observed. The objective reality of an observer being separate from that which he is observing does not exist because the observer is quantum entangled with what he is observing. Wave-particle duality is when a particle that is confined to a very small volume of space is simultaneously a wave spread out across a large region of space.

The idea that anyone could enter into a sound and experience quantum behaviors was absurd to physicists at the time. The same quantum behaviors that shocked physicists have been explored and documented by musicians and sages for thousands of years. There is an old saying that says, "The scientists climbed the mountain and found the musicians and mystics waiting." Scientists are trained to objectively observe behaviors, whereas musicians and sages learn to enter into a sound and directly experience its "teaching." When our conscious awareness is located inside a sound, our inner experience of reality changes and comes into resonance with the vibration of the sound. When this happens, we enter a state of "mindful listening" in which we have experiences that are similar to the quantum behaviors described by physicists. The composer, Terry Riley, wrote in the Trinity of Eternal Music:

In the sound, we become aware of its subatomic structure, an eternal dance of particles—playful, recombining, recreating, and dissolving.[8]

Today, the quantum understanding of the universe is slowly maturing into a science that includes both objective science and subjective inner experience. The parallels between mysticism and quantum physics have been reported by many quantum physicists for many years. The most well known discussion of quantum mysticism is Fritjof Capra's book *The Tao of Physics*. The parallels between mysticism and quantum mechanics have been discussed since the beginning of quantum science.[9] The question that is repeatedly asked by physicists is, "How did they know about quantum behaviors without our modern scientific instruments and knowledge." The answer is simple: *They knew because they listened.*

PART 2
Mindful Listening

Listening is nothing less than our 'royal route' to the Divine. —Dr. Alfred A. Tomatis

Mindful listening is the most fundamental skill that is required to have for sound healers. Mindful listening is the ability to be aware of and to consciously enter into sounds by focusing one's awareness on the present moment, while calmly accepting thoughts, feelings, and body sensations without judgment. If you are mindfully listening to a sound, you are practicing sound healing in its simplest form since *all sounds have the potential to be healing*. In order to embody this statement, it is necessary to learn to safely listen to sounds without constricting one's body. Safety, in the broadest sense, is any sound the listener can listen to without amygdala activation.

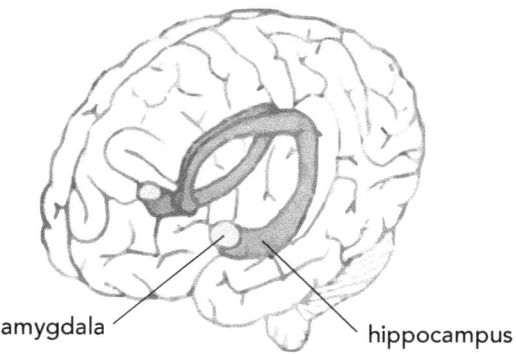

The amygdalae are two almond shaped structures located on the left and right temporal lobes near the interior horns of the lateral ventricles of the brain. The amygdalae can be visualized as the gatekeepers of the limbic system. They are constantly scanning for danger signals and threats. The amygdalae activate when we perceive something that is not safe or is threatening to us. When this happens our body contracts, our heart and breathing rate increases, and we gear up for fight or flight conditions. When the amygdalae are stimulated electrically, animals respond with aggression. If the amygdalae are removed, animals get very tame and no longer respond to things that would have caused fear before.[1]

The amygdalae receive sounds through two primary pathways, the hearing pathway and the listening pathway. The hearing pathway is necessary for survival and the listening pathway is necessary for health and well-being. The hearing pathway bypasses the auditory cortex and goes directly to the amygdalae via the auditory thalamus. This pathway exists so that we can react instinctively and quickly to potential dangers. For example, if we hear the sound of a siren or another loud noise, we immediately react and become alert. The listening pathway passes through the auditory cortex to the thalamus without activating the amygdalae. It is slower and transmits many subtle levels of sound. The listening pathway is a mindful experience of focused awareness of sound that has the potential to transform the act of listening into a harmonious experience with profound biological and psychological effects.

Like many other pathways in our body, the hearing and listening pathways are integrated by balance. For example, if you hear a siren, your amygdalae immediately activate and a signal is sent via your limbic system, which activates your sympathetic nervous system, and you immediately become more alert. It is important at this point to evaluate the sound and, if you are in danger, to take action. However, if you are not in danger, you can learn to shift into the listening pathway. The ability to shift from hearing to listening is the key to understanding and perceiving a sonic universe. One of my favorite sound experiences is listening to the high-pitched "beep beep" of electric carts moving through a crowded airport. This "beep beep" is designed to activate the amygdalae to wake people up from their airport trance and to move out of the way.

If you are napping or relaxing in a chair when the cart goes by, it can be an intrusive experience. You can instantly go from a dream-like state to a fully alert or on-guard state. At this point you have a choice. You can either get angry at the airport for using high-pitched "beep beep" carts that mess up your relaxation. However, this will not do you much good and you may even walk around angry all day. This is your choice. The other option is a method I developed to convert disruptive environmental sounds into positive energy experiences called environmental toning.[2] It begins with making sure you are safe and that the "beep beep" sound is just a sound passing by. Next you can shift back into your dream state, be with the sound, and mindfully listen to it like a mantra or music concert.

I was sitting comfortably in the airport waiting for my flight to Switzerland. I heard a high-pitched beep-beep sound in the distance slowly coming closer.

I knew the electric cart was working its way through the crowd. I closed my eyes and focused on the sound. As the sound came closer, it was no longer the sound of an electric cart. I allowed myself to "get inside" the sound and to feel each high-pitched beep in a different area of my brain. Suddenly, I was walking along Jacob's Way on a pilgrimage with Paracelsus, the alchemic Swiss physician and grandfather of modern medicine. With each beep he smiled and instructed me on the awakening of "the seven brain stars" as an integral part of alchemic consciousness. This lesson in alternate time took many hours but was over quickly in real time as the high-pitched beep-beep disappeared into the distance of the airport.

When I got to Switzerland, my friends Andreas and Brigitta, without knowledge of my airport sound experience, took me to Einsiedeln where Paracelsus had lived. We walked Jacob's Way near where he had tended to passing pilgrims. We visited the ancient Benedictine Abbey of Einsiedeln where we drank from the water of seven different springs and paid homage to the Black Madonna with the sound of Gregorian chants in the background.

Everyone's listening range and amygdala activations are based on their own unique associations with different sounds.

When I worked at Bellevue Psychiatric, we took the children from the ward to the Frost Valley YMCA camp in the Catskills for a nature adventure. They were having a great time swimming and playing in the woods. I remember thinking that this was great and what a wonderful experience this must be for these disturbed inner-city children. That evening we had a campfire and told stories. They were "happy campers." We all went back to our cabins and made sure the kids were safely in bed. I woke up at 1 AM to the sound of children talking. I asked them what was going on. I thought they were probably excited and couldn't sleep. They said they were afraid. I thought maybe they were afraid of a bear or of other wild animals. One of the campfire skits had involved bears in the woods. They said, "It's all those noises." I thought, "What can they be talking about. There are no car horns or sirens out here. It's so quiet and still." At that moment, the camp counselor came into the cabin carrying a ghetto blaster. He said, "Don't worry, everything will be all right." At the sight of the ghetto blaster the children visibly relaxed. He put it on a table and turned it on. Suddenly, I heard the sounds of cars, buses, motorcycles, and sirens that the counselor had recorded in New York City. The children were asleep within five minutes. The camp counselor looked at me and said, "It never fails."

Beliefs can also create amygdala activation. I presented tuning forks to a group of skeptical worshippers at a Pentecost revival. They were concerned that tuning forks could be the "devil's instruments." It was clear that their amygdalae were activated just thinking about the sound of the tuning forks. I told them that the tuning forks were the same as the tuning forks musicians use to tune their instruments. The tension in their facial muscles immediately began to relax. Next, I said that the tuning forks were special because they were tuned to the same pitches as church bells. Everyone got excited and we spent an hour playing "tuning fork church bells."

We are not always aware of it, but sound carries complex social and cultural signals that can become beliefs. For example, in the city of Boston the sound of jackhammers were listed as a top "noise complaint." Sound engineers recommended the simple solution of mufflers to reduce jackhammer noise. The muffled jackhammers worked; however, the jackhammer operators thought that the muffled jackhammer didn't work because they had lost the sound indicator that they had associated with "working right."[3]

Many people driving electric cars tend to drive too fast because they don't have the feedback of the sound of the engine. Many pedestrians are also used to listening for engine sounds before crossing a street. Rather then adjust to a new silence, much like jackhammer workers adjusting to muffled jackhammers, laws are being made to add engine sounds to electric cars. What is more interesting is our attachments to and beliefs about sound.

Here are some blog comments that people have made about car sounds or the lack of car sounds:[4]

> "That is ridiculous, the sound of an electric car. I like the gasoline sound. These guys need to make a realistic car sound, not a computer game sound."

> "Great, just as we're about to get rid of cars' noise pollution, the retards add it back. While you're at it, add a furnace so it can produce some smoke too, will ya?!"

> "A fast car without sound is like a party without music."

"I hope in the future every company adopts this fake sound system. A soundless car is a joyless car."

"They should have kept it silent because it would make it a unique selling proposition. I want a quiet car not a loud one."

"This is a great idea. I'd never buy a silent car."

"This is stupid!"

IDENTIFYING AMYGDALA ACTIVATION

A sound healer should know how to identify the signs of amygdala activation both in themselves and in the people they work with. The most common ways of evaluating amygdala activation in oneself or in others are:

- feeling or observing a tightening and/or body constriction
- having thoughts about a sound or expressing displeasure with a sound, for example, "I don't like that sound" or "That sound reminds me of something that's not good"
- becoming aware that you are having negative beliefs or associations with a certain sound or music

MINDFUL LISTENING EXERCISES

During mindful listening, our conscious awareness is immersed in a sound and our rational-objective mind becomes a passive observer. When engaged in mindful listening, the mind becomes silent, physiological activity slows, awareness increases, and we create a new perspective and relationship to life events. The practice of mindful listening helps us to slow down and to be present and aware in each moment with sounds. The following exercises are designed to open our ears to different sounds in a non-judgmental way.

Mindful Breathing

Mindful breathing is essential to all mindful listening practices. When the amygdala is activated we tend to take short, fast, and shallow breaths or long, forced breaths. Research has shown that breathing rhythmically will disengage the amygdala response, relax our emotions, and open us to mindful listening. The American Institute of Stress recommends a ten-minute breath cycle with a 5-second inhalation and then a 5-second exhalation.[5] Research, in general, suggests that relaxed, long rhythmic breathing will have a positive effect.[6]

Begin by rhythmically breathing in and out in slow deep breaths through your nose. Gradually move to a ten to twelve breath cycle. One breath cycle should last for approximately 10 to 12 seconds.

1. During your breathing, let go of your thoughts, i.e., let go of things you have to do later today or pending projects that need your attention. Simply let your thoughts effortlessly come and go, don't pay attention to them, and don't try to stop or control them.
2. Let your breathing rhythm become automatic and focus your sense of awareness on its sound and rhythm.

Environmental Mindful Listening: Transforming Noise into Music

Environmental sounds are sounds which are audible within our immediate sonic environment, such as horns, train whistles, screeching tires, planes, machinery, water dripping, children playing, and phones ringing. These sounds are often labeled as "noise." These environmental sounds can be transformed into exciting concerts when listened to in a mindful way. Your life will become easier, more vibrant, magical, and childlike.

1. When you become aware of an environmental sound, first and foremost, mindfully breathe.
2. Make sure you are safe.
3. Suspend your mental attitudes, relax, and open yourself to the sound.
4. If you notice signs of amygdala activation, let your voice imitate the sound and let your body be moved by the sound.

EXAMPLES

1. A car horn suddenly honks and you notice that your body tightens (amygdala activation). Your mind is judging the driver and your emotions are held in. The situation is similar to touching a hot stove without letting out a sound. Therefore, instead of holding the sound, jump back from the car, allowing your body to unwind the tension and then allow your voice to make a loud spontaneous "honk!"

2. You are sitting quietly in a park mindfully listening to the birds chirping. Across the street a garbage truck is picking up trash and making lots of "noise." Your body tightens and you angrily judge the city for allowing these sounds that interrupt the peacefulness of the park. In this moment you have a conscious choice. You can continue being angry and judgmental, or you can see it as an opportunity to practice mindful listening. If you choose mindful listening relax, enter into mindful breathing and enter into the sound. Close your eyes and let go of your beliefs, thoughts, and emotions about the sound. Relax your jaw and let your whole body vibrate with the sound until you become the sound.

3. I was walking down a New York City street with two friends. The trucks were idling and the workers were unloading crates. Something about the sound of the trucks caught my attention. I asked my friends if they would be willing to watch out for my safety while I sat down and listened more deeply to the sound of the trucks.

    I sat in a safe spot and allowed my body and mind to relax. The sounds of the trucks formed a distinct rhythm: da-da-daa – boom, da-da dada-daa – boom, etc. I let myself mindfully listen to the sound. At first I imagined being a teenager in Indiana and going to the drag races and listening to the sound of the engines idle. I remembered how these sounds had fascinated me and I would imagine driving a race car.

    Listening even deeper to the sound, I felt myself "move inside" its pulse. My imagination and memories were still present, but they were now accompanied by a new sensation. A profound quietness or stillness came over me and, for a moment, I experienced myself as just being the sound. Then there was a shift in my awareness and I found myself with a group of Aborigine chanters. They were communicating a message about "dream time." I was just as clearly with them as I had been with the truck just a few minutes ago. I allowed myself to absorb their message. An old man signaled that they must move on into the desert. I understood. At that moment, a friend tapped me on the shoulder and I came back to the reality of a New York City street. The truck was driving away.

Mindfully Listening to Music

1. Choose a piece of music to listen to. Be aware that different music can possibly bring up thoughts, emotions, and beliefs via hippocampus activation. How this happens is discussed in the Sound and Values Visualization section. Therefore, it is best to begin with a neutral piece of music that consists of instrumental sounds. Lyrics tend to distract and focus the mind on memories and past emotions. One can learn to mindfully listen to lyrics, but it is best to learn with instrumental music.

2. Mindfully breathe and relax.

3. Give yourself permission to listen to the music without having to check your email, text messages, etc. Notice your body and release any tensions. It may help to close your eyes to tune out any visual distractions. Use headphones or earbuds if that helps you to focus or to shut out external noise. During listening, notice the pace of the music, the sounds of the different instruments, or the shifts in volume. Notice if you're more aware of a certain part of your body as you listen. Let any thoughts that may come up pass through your awareness, and then gently bring yourself back to the sounds of the music.

The Deepest Level of Mindful Listening

The deepest level of listening is silence. The center of all sound is silence. All sounds arise from and lead back to silence. Mindful listening is the art of discovering silence. Silence is the key to many adventures the world of sound has to offer. Through silence we are truly safe and free. We know the beginning and we know the end.[7]

Silence

PART 3

Be Like a Child

Verily I say unto you, except ye be converted, and become as little children, ye shall not enter into the kingdom of heaven. —Matthew 18:3

When our conscious awareness is mindfully focused on a sound, we simultaneously entrain with that sound. As our inner experience of reality comes into entrainment with the sound, our consciousness and inner awareness expands and our rational-objective mind becomes an aware yet passive observer. Musicians sometimes refer to this as "getting inside a sound." When we are inside a sound, the sound will become music to our ears. From this perspective, music is the appreciation of sound, therefore all sounds have the potential to be music. The composer John Cage says:

> "If you develop an ear for sounds that are musical, it is like developing an ego. You begin to refuse sounds that are not musical and that way cut yourself off from a good deal of experience."[1]

John Cage was speaking about the division of the universe into musical sounds and non-musical sounds by musicians. However, one person's "noise" can be another person's music. Visually, it is said that "Beauty is in the eye of the beholder." The same is true for sound: "Music is sound in the ear of the beholder." John Cage made this point in a talk in 1937 when he said that disagreements among musicians in the immediate future will be between noise and so-called musical sounds.

All sounds and all music can be potentially healing. However, this is like saying all yoga students should be able to touch their toes. The realization of all sounds as healing is a process that requires a willingness to expand our ability to play and to mindfully listen to different sounds without amygdala activation. Just as stretching our body will give us more physical flexibility, "stretching our ears" will increase our neural flexibility, or what is called neural plasticity. The general rule is **the more sounds you listen to and explore, the more open you will be to new ideas, new possibilities, and new opportunities. Opening ourselves to the beauty and mystery of all sound and its expressions is a fundamental**

principle of sound healing. As we appreciate more sounds, we change from the inside out, because we are experiencing life in a new way. This is very different than trying to change someone from the outside in by having a rational conversation or argument.

Sound healers have a lot in common with composers who are always experimenting with and exploring new sounds. For example, I am always on the lookout for hollow tubes, wooden blocks, sheets of metal, pipes, and cans that make sounds. I often play two flexible plumbing pipes I found in a hardware store bin on Canal Street in New York City in 1974 for twenty-five cents each. They create beautiful high-pitched flute-like sounds that are in resonance with the overtone series.

Found Flute Pipes

John Beaulieu Playing Found Flute Pipes in Zug, Switzerland, Concert

There are many examples of composers altering traditional instruments to make new sounds. In 1922, the composer Henry Cowell created the "string piano." He instructed pianists to reach inside the piano and to pluck, sweep, scrape, thump, and play the strings with their fingers rather then using the keyboard.[2] Inspired by Cowell's expanded piano sounds, John Cage invented the prepared piano. To prepare a piano, Cage put nuts, bolts, and pieces of rubber between and entwined around the strings. This created duller sounds, percussive sounds, and bell-like tones that were very different than normal piano sounds.[3]

The difference between a sound healer and a music composer can be demonstrated by the playing of a piano lyre. When Cowell and Cage first experimented with the piano lyre, they were "playing with sounds." This is the "spirit" of a sound healer. Cowell and Cage created compositions to which they gave specific instructions on how to make the sounds they specified during performances. A sound healer can explore the sounds of a piano lyre much like Cowell and Cage did. They are like children playing with sounds without regard for the original use of the piano as a musical instrument. Cowell and Cage created notated music compositions in their time, while a sound healer learns and plays the piano lyre to create improvised sounds coming from a healing intuitive connection with their listeners.

Listening to sounds like a child requires that you remember innocently playing and listening to sounds beyond adult associations and beliefs. If I take a sound healing student to my piano and ask them to make sounds like a child, they usually become nervous and their body tightens. They either think they have to play music and/or they recall memories of piano lessons. I own an old Baldwin studio piano that I stored in my garage. It got infested with mice and I was told that it was beyond repair. Rather then throw out the piano, I took it apart and threw out everything but the lyre and the sounding board. I polished it up and created a completely new "lyre instrument."

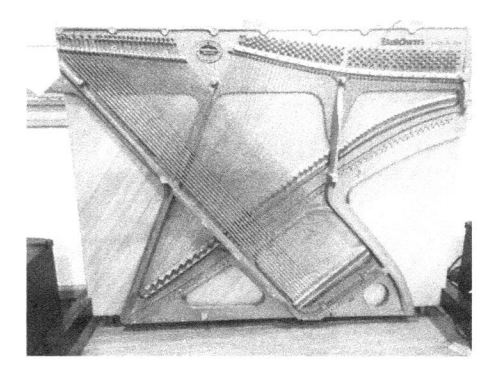

During my sound healing classes, students have no associations or experiences with piano lyres. Many don't even know what they are looking at. It doesn't matter; they know it makes sounds. They experiment making sounds with the lyre by using fingers, strikers, picks, and rubber balls. I observe the smiles on their faces and the glow in their eyes when they discover new sounds. They are like little children playing because the lyre is not associated with prior memories or beliefs.

The following is an example of playing with sounds from Iannis Xenakis, one of the greatest composers of the 20th century. He began his career as an architect and engineer. During WWII, he served in the Greek resistance and afterwards in the Greek struggle against communism. During this time, he experienced the healing effects of music and had a vision that he should become a composer. He wasn't a musician and had no musical training. His strength was in mathematics and he loved sound and music. He wrote *Formalized Music: Thought and Mathematics in Composition* in which he brought the Greek tradition of mathematics and music into the 20th century.[4]

In order to study Formalized Music composition, I learned calculus, probability theory, set theory, game theory, statistical analysis, and topology. One very late evening at the Indiana University Music Lab, when it was rare for anyone to be in the building, I was playing the Moog synthesizer when I heard very strange sounds coming from somewhere in the building. I was fascinated and so followed the sounds to the room from which they were coming. I opened the door and, to my surprise, saw Iannis Xenakis holding a live microphone by a cable over a charcoal grill. As he lowered the microphone closer to the hot coals, a very strange crackling sound came through the speakers. Xenakis looked up, his eyes glowing, and said to me, "Did you hear that?"

I got home around 3 AM that evening and went to sleep thinking about my many hours of studying mathematics and the absurdity of what I had just witnessed. Suddenly in visionary flash, "I got it!"

Let go of your rational logical mind and listen to sound like a child.

Learning to listen like a child can be challenging. As we have seen in the previous part, the amygdalae can become activated based on our past experiences and beliefs. When a person sits at a piano keyboard or holds a guitar, they tend to think they should play a style of music. As electronic music instruments began to become available, such as the theremin in the late 1920's, rather then explore new sound potentials, musicians used theremins to recreate 18th and 19th century music.

A theremin is the only musical instrument that is played without physical contact. Instead, players move their hands around two metal antennae. One antennae controls pitch and the other controls volume. It is named after the Soviet inventor Léon Theremin who patented the device in 1928.

The new electronic sounds were not yet "real music," just as muffled jackhammers were not "real jackhammers." The Moog synthesizer gained a lot of popularity after the release of the album *Switched-On Bach* in 1968, although it was met with negative feedback by Bach purists and classical music traditionalists. The album went on to sell over two million copies. Its popularity was primarily with young listeners who have the aural flexibility to appreciate something different.

LISTENING LIKE A CHILD GUIDED MEDITATION

This guided meditation can be done alone, with a friend, or with a group. It is suggested that you have safe ambient sounds playing in the background and/or someone tapping tuning forks or playing instruments such as sounding bowls. If you do it alone, you can listen and read the meditation. If you do it with a friend or group, then another person can read the meditation. The recommended BioSonic tuning forks to tap are C256, F, G, and A or just C256 and G in the Solar Harmonic Spectrum. If you are listening to a tuning fork recording in the background, it is recommended that you listen to *Calendula*,[5] *Apollo's Lyre*,[6] or the ambient sounds of crystal or metal bowls.

> Listen to the sounds and let thoughts about your day come and go effortlessly and spontaneously into the sound. Imagine that your thoughts are like trees in a forest filled with leaves. Look out across an open meadow, just beyond the trees where the sun is shining. You can see the light and you begin moving towards it. With each step, you pass tree after tree, thought after thought. As you get closer to the meadow, you become aware of thoughts that were given to you in childhood about what is and what is not music. Take a moment to allow yourself to review these thoughts.

Go from experience to experience and make peace with each thought, each person, until you come to the moment where you choose to make sounds, some musical and some not musical, some sounds OK and other sounds noise.

Take a moment to be with your inner child. Notice that just beyond this place is a meadow filled with sunlight and wildflowers, a place where you can listen to the music of flapping butterfly wings, the sound of buzzing bees, and you can see the opening of flowers to the sun. Here is a place where all sounds arise and return to silence within the center of sunbeams.

Become aware of who told you to make the distinction between sound and music, and between sound and noise. Who told you that there are good sounds and bad sounds. How was this belief passed on to you? What is the feeling of this belief being passed on to you?

If you have any thoughts of playing "wrong sounds," let them go. In the meadow there are no wrong sounds, only sounds that dance in the sunlight, creating colored wildflower patterns. Say to your child self, "It's OK for me to enjoy listening to and playing sounds without thinking about whether the sounds are the 'right sounds' or are music." Just be with the sounds and they will take you to magical places.

As you return to your normal consciousness, having just reconnected with your love of sound, begin to enjoy now what composers and musicians have shared about learning to play in the meadow of sound.

A 'mistake' is beside the point, for once anything happens it authentically is. —John Cage

It's not the note you play that's the wrong note—it's the note you play afterwards that makes it right or wrong. Do not fear mistakes. There are none. —Miles Davis

There's no such thing as a wrong note. —Art Tatum

Anything played wrong twice in a row is the beginning of an arrangement. —Frank Zappa

I frequently hear music in the heart of noise. —George Gershwin

PART 4
Listen Like a Scientist

The only acceptable point of view appears to be the one that recognizes both sides of reality—the quantitative and the qualitative, the physical and the psychical—as compatible with each other, and can embrace them simultaneously.

—Wolfgang Pauli, winner of the Nobel Prize in physics[1]

Sound is our real teacher. To meet and respect our teacher, sound healers need to listen to and explore sound in a systematic way. The systematic exploration of sound and music requires mindful listening and a form of inner experience research called phenomenological research. The phenomenological research approach relies solely on one's personal experience of sound.[2] In this kind of research, we learn to examine our listening experiences of sound from the inside out. This stands in stark contrast to the reductionist model which studies sound from the outside in. Reductionist scientists do everything possible not to be in relationship with what they are researching, even though the quantum science on this is very clear: We can never be separate from that which we are observing. When we know a sound from the inside out and integrate our "knowing experience" with modern reductionist research, we will always arrive at a holistic understanding.

Listening and objective accountability should be seen as a continuum that is balanced like a teeter-totter. There are situations where we need to be more focused on objective accountability, and other situations where we need to be more focused on letting go and listening. Ultimately, objective accountability and listening are happening simultaneously all the time within the greater whole. It is possible to be more objective and at the same time receive non-objective listening information. It is possible as well to be immersed in a listening experience and at the same time be aware of objective reality.

Dr. Fritz Perls, the founder of Gestalt Therapy, said, "Lose your mind and come to your senses."[3] When we let go of our rational mind, we enter a vibrational universe of flowing sensory experiences that exist beyond the meanings we give to normal life experiences. At first it would appear that letting go of our mind means letting go of objective accountability. This was the fear of the non-logical quantum behaviors expressed by many physicists. How could one maintain objectivity or any form of accountability within a quantum field that defies all laws of logic?

As it turns out, we let go of our objective mind all in time in situations that require accountability. Airplane pilots are responsible for the safety of all the passengers on their airplane. They undergo rigorous objective training to be able to handle all situations. They understand that there is a safe time to put their plane on autopilot, sit back, and let the plane fly itself. During this time, they are always monitoring, and if anything happens that requires their attention they will instantly take back control. This is an experience we know as pilots of our cars. When we drive down the interstate, we can put our cars on cruise control and enter a state of highway hypnosis. If we see a brake light, our objective mind instantly returns and we take back our control.

I sometimes think that researchers and scientists can become too immersed in objective reality, whereas musicians and composers can become too immersed in the quantum-like behaviors of sonic realities. We don't enter into sounds to get high or to have "cool experiences." We enter into sounds with the following intention: To safely explore and learn from sounds in order to expand our consciousness. When exploring sounds, it's important for us to have our feet on the ground and our head in the clouds. I learned this many years ago as a music student at Indiana University.

> In 1971, I finally got access to the Moog Synthesizer Lab at Indiana University. In those days, the Moog Synthesizer was revolutionary and took up a whole room. I had taken a course on the Moog, but the only time I could get in for private experimentation was at 1 AM in the morning. I played with the sounds of the Moog for a least 2 hours. I became lost in the sounds. When I left the lab I had no idea who I was, what time it was, or what reality I was in.
>
> Fortunately, two of my friends and fellow musicians were walking by after a party. They knew I might be in the lab and decided to stop by. When they saw me, they realized I couldn't focus. I tried to talk and my voice just echoed into waves. I was ready to fly on the light waves of street lights. My friends grabbed me and took me to White Castle, an all-night hamburger restaurant. I was a vegetarian but, to help me, they ordered lots of 25-cent

mini hamburgers. I looked at them and something popped. Just like that I came back to "normal reality." We laughed and laughed. Thank you, White Castle!

From the time I was 3 years old, I had spent my life entering into and journeying with sound. The sounds of the Moog synthesizer took me to another place. It was completely unexpected and I learned a lot that evening. To sum it up:

Sound is powerful! Sound mimics and resonates with the vibratory nature of reality: Be responsible and know what you are doing.

SOUND JOURNALING

A Sound Exploration Journal is a phenomenological method of inquiry that creates a structure in which you can observe, record, and learn from your sound explorations. One purpose of keeping a sound journal is to develop an objective internal witness during your sound experiences. For example, you may be somewhere and a sound catches your attention. While you are immersed in the listening experience, your internal witness should be present and/or available, but in no way should it interfere with the experience.

Because the experiences that can happen during sound explorations may not seem logical and may appear in many different ways, the structure of your journal and the way you record your experiences should be creative. Your Sound Exploration Journal can take on any form that is congruent with your experiences. Here are some examples for you to explore in your Sound Exploration Journal:

1. Record your experiences in a logical order.
2. Draw your experiences in lines, scribbles, or pictures.
3. Discover pictures in a magazine or on the internet that remind you of your experience. Cut them out or print them and paste into your journal.
4. Place one experience within another, i.e., write something inside of scribbles and also put a cut-out picture nearby.
5. Draw arrows between experiences or use color codes to show relationships between experiences.

6. Record your sound experiences in the form of information downloads. An information download is always happening while listening. It's possible to be aware of "just listening" during the listening process and, over a period of time, you may become aware of information the source sound is giving you. Information downloads can take place in dreams, in trance states like driving down the road or flying, during meditation, or actually anytime. If you get nothing, then write nothing or something that represents nothing. Sometimes it takes time for sound experiences to come into our conscious awareness and they often come through unexpected and non-logical channels. For example:

   a. *Sound and dreams:* During your sound exploration, you may not be aware of experiencing anything. Maybe you just spaced out or went to sleep. However, that night you may have a dream that is clearly related to your sound experience. You may not understand this logically, but you "just know." The dream should be in your journal with that sound experience.

   b. *Sound and body feelings:* During your sound exploration, you experience a "feeling" in your body. Later that day or even weeks or months later, you experience a similar feeling that reminds you of your sound experience. Go back into your journal and document this with the recording of your original sound experience.

   c. *Sound memories:* During your sound exploration, past memories surface and, that day or during the next week or month, someone contacts you from your past, associated with the memory, or someone you would not expect mentions something to do with your memory. Go back to your journal and document this. It's possible to be inside a sound and have a memory that you have been there before. This is sometimes referred to as déjà vu.

The following is a summary of areas of experience that may or may not happen during a listening experience. They are ideas for questions you may want to ask yourself after your listening experience. Feel free to change them or to make up your own. Use your Sound Exploration Journal to explore the following and document each one.

1. *Thoughts about the sound:* Are you aware of any preconceived thoughts about the sound? These thoughts can include opinions, previous experiences, beliefs, and stories others may have told you.

2. *Body sensations:* What do you feel in your body while listening, i.e., any constrictions, openings, relaxations? In what areas do you feel tingles, shivers, pops, etc.?

3. *Feeling tone:* What is the general quality of the feeling of the sound in your body, i.e., soft, harsh, even, smooth, fast, slow, expansive, etc.?

4. *Spontaneous thoughts:* Are you aware of thoughts that spontaneously appear as you listen, i.e., I like this sound, I do not like this sound, this sound is good for these types of people? This sound can be used in these different ways:

    a. *Temperature:* Do you sense the sound as a warm sound or a cold sound?

    b. *Colors:* Does the sound have a color or colors?

    c. *Emotions:* Do you experience emotions as you listen to the sound, such as sadness, happiness, anger, excitement, grief, courage, desire, joy, etc.?

    d. *Associations:* Are you aware of any associations you might have with the sound such as childhood memories and beliefs?

TUNING FORKS AND PHENOMENOLOGICAL RESEARCH

Tuning forks lend themselves to systematic phenomenological research because:

1. They can be tuned and labeled for a specific frequency and/or interval.

2. Researchers using them will have a common tuning baseline from which to record, discuss, and correlate experiences.

3. They are easy to learn and use. Anyone can learn to use them and record their experiences. This takes science out of reductionist labs filled with expensive equipment and empowers individuals to learn about and conduct their own studies.

4. The precise tuning makes them available for reductionist experimental designs.

In my teaching studio, I have crystal bowls, metal bowls, crystal pyramids, element chimes, Peruvian Whistles, rain sticks, flutes, drums, bells, toy instruments, synthesizers, a piano lyre, a 1914 Steinway grand, and much more. However, what students notice the most when they visit is over one thousand differently tuned tuning forks in stands and hanging on the walls. I first began phenomenological experimenting with two tuning forks in 1974 in an anechoic chamber in a New York University psychology research lab.[4] Fortunately, during my doctoral studies, the importance of a phenomenological approach to research was emphasized by my professors. I immediately realized the research advantages of using tuning forks. Over the years, I have experimented with hundreds of different tunings.

At first I was exploring the effects of specific sounds on myself. For example, I would isolate different intervals such as C & G (Perfect Fifth), C & F (Perfect Fourth), and I would listen to the interval. I got very creative with my explorations. I bought a Bone Fone from Radio Shack through which I played recorded tuning fork intervals. I would spend a week or more "Bone Foning" an interval. I kept Bone Fone under my clothing so no one could see it and kept the volume just low enough to sense the vibrations through my skin but not interfere with my daily activities. Here is what the ads for Bone Fone said in 1979: "You're standing in an open field. Suddenly there's music in all directions. Your bones resonate as if you're listening to beautiful music in front of a powerful home stereo system. But there's no radio in sight, and nobody else hears what you do. Who wouldn't want this experience?"[5]

I also slept between two speakers and played the interval I was exploring all night. I kept the volume at an almost subliminal level and combined my sound exploration journal with my dream journal.

I soon became very interested in microtonal overtones. To systematically explore each overtone, I created tuning forks to isolate different overtones and their intervals.

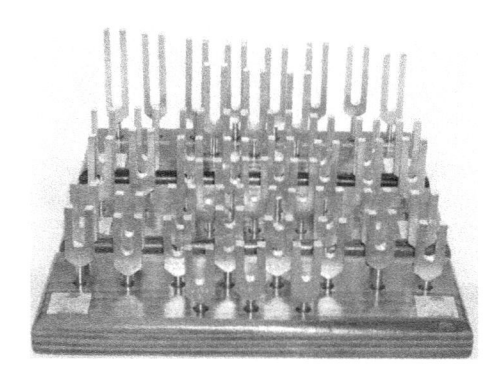

The above photo is a set of fifty overtone tuning forks created for phenomenological experimentation. The overtone ratios are converted into microtonal music intervals through mathematics. The equation used to do this is: Frequency/X = ratio, i.e., 1:2, 2:3, …6/7, or any variation of harmonic relationships, i.e., 4:9.

During this time, I became inspired by my experiences with microtonal Peruvian Whistling Vessels (*shown above*). When investigating the high-pitched overtone sound of Peruvian Whistling Vessels, physiologists at the Franklin Institute in Germany reported definitive bodily changes in heart rate, blood pressure, respiration, and basal metabolism.[6]

In 1990, I commissioned a set of 50 overtone whistles (*photo at top of next page*) from master whistle maker Kathy Tighes. My intention was to create an overtone sound immersion experience for people in which each overtone was sounded at an equal volume. I wanted to give the listeners the experience of different overtone groupings or what I call harmonic networks. I created a set of numbered cards for each overtone, whereby I could signal a whistle blower to be on or off. I used an intuitive pendulum divination system during each concert to control overtone whistle networks.

I recorded my tuning fork overtone immersions on my CD *Calendula*, Track 16: "Angelic Conversations."[7] I recorded my overtone whistle immersions on my CD *Spirit Whistles*.[8]

Overtone Whistles

You need to be both systematic and creative in your explorations of sound. Have fun and share your experiences with others. It's not necessary to focus on a predetermined outcome. Enjoy the unexpected and something positive will happen. For example, during my first sound interval explorations, I developed the ability to "sense" a person's tuning. I started creating tuning fork concerts for my friends based on their tuning. There came a point when I could be with a client and I would know which tuning forks were best for them.

One can never underestimate what a group of people can accomplish by having fun and sharing experiences. For example, Foldit is an online puzzle game about protein folding developed by the University of Washington Department of Biochemistry. The researchers realized that people have intuitive spatial reasoning skills that computers are not good at. They created a game to bring people together to solve protein folding problems that had stumped scientists for many years. In just three weeks, ordinary people solved a molecular puzzle concerning AIDS protein folding.[9]

PART 5
Sound Stories to Warm Your Ears

EARL'S AUCTION

When I was a teenager in Indiana I worked for Earl's Auction. Earl was a Midwestern farmer and auctioneer. Earl realized I had a talent for "calling auctions" and made me an assistant auctioneer. In those days, a Midwestern auction was 25% an auction and 75% a social event. The 25% was auctioning off houses, cars, and farm equipment. I was part of the 75%. I auctioned buttons, old clothes, records, hardware, household utensils, old lamps, rugs, etc. Nothing was too small for Earl to auction off.

One of my favorite strategies was to tell a brief story of what I was about to auction. In my Midwestern grammar I would say, "This here jar of buttons is Ruby's special collection that she always shared with her knitting group friends."

Of course, someone from Ruby's knitting group would be in the audience and say, "I got a button from Ruby all those years ago." And that would be my cue to start my auction chanting in a fast, high, nasal rhythm. The audience, which was mostly made up of farmers looking for some weekend excitement, would light up. I would watch their facial expressions and modify my voice quality to get a reaction. There is nothing like seeing a stern farmer's face come alive and his hand going up to make a bid. Earl used to come to hear me auction and one day he complimented me. He said, "Boy, you sure do get 'em excited over nothin'."

Many years later, I heard a talk by the composer John Cage. He said, "I have nothing to say and I am saying it." I thought about Earl's compliment all those years ago. I thought I had nothing to sell and I was selling it. To this day, I think about how amazing it is that people get so excited about another person's things and, rather than let them become junk, they would give them a new story. Sooner or later though, everything auctioned is destined to become "no-thing."

INDIANA HOG CALLING

When I was 26 years old and living in New York City my friend Bob called me from Indiana and said, "John, you have to come back to hear this hog caller." I said, "Bob, you want me to go all the way back to Indiana just to hear a hog caller?"

He said in an excited way, which was completely against his character, "I guarantee you ain't heard nothin' like this." Now, I had done my share of hog calling, and had seen some really good hog callers in action. However, I was living in New York City and hog calling was not a priority for me at the time. What really got to me was the excitement in Bob's voice. My intuition kicked in and I said, "OK!"

The hog caller, to my surprise, was a woman from China, billed as the World Champion Hog Caller. She had come all the way from China to show off her hog calling skills at the Indiana State Fair to a bunch of farmers. I remember sitting in the audience with my friend Bob, surrounded by farmers. The arena was packed with the best Indiana Hog Callers, and I can say from personal experience that farmers in Indiana know how to call hogs. We were looking down at ten hogs snorting around in an arena. The goal was to get them to move into a pen at the other side of the arena as fast as possible.

This very small woman walked into the arena with some pots and pans hanging around her neck. I was told that she didn't speak English. She went to the middle of the arena and looked around at all the farmers. We were all elbowing each other and joking that she couldn't possibly call hogs. I watched her move to about 20 feet from the hogs. The hogs were just snorting and having a good time, completely unaware of her. At this point, the judge signaled her and the hog calling competition began.

Everything about her changed when she focused on the hogs; it was as if she had become a different person. Everyone became very quiet. She rattled her pots and pans and the hogs started to snort and look up. There was a brief silence before she took a deep breath and then she let out a high-pitched, piercing "soo eee eeeeeeeeee!"

I felt the sound move through my cranium like a bolt of lightening. The whole world changed and it was as though I were watching from another dimension. The hogs stopped in their tracks and just stared into space. She screeched again and the hogs went crazy. She hit her pots and pans and moved the hogs towards the pen with what seemed to be her will. They all started to run and she let out yet another "soo eee eeeeeeeee."

The farmers were on their feet. It was as if the sound levitated them out of their seats. I was standing up and couldn't remember standing up. I was ecstatic. I didn't know a hog could run that fast. I had never heard anything like that sound and my head was spinning. The farmers were either speechless or were exclaiming, "What the hell!" The judge finally announced that she had broken the Indiana hog calling record.

FLAGPOLE CONCERT

When I was a graduate student at Indiana University, my son was 4 years old and we were members of one of the first co-op day care centers. One day, while on my shift with the children, I went with two other parents and twelve children on an adventure to a football field. It was a spring day and the field was completely empty. We took the children to the middle of the field and turned them loose. They all ran in different directions. Suddenly a strong wind came up. We heard the ropes on the flag poles striking the poles. It was an amazing sound. I counted twelve flag poles spaced at equal intervals around the stadium. As the wind came through, they sounded like wind chimes. Each pole made its own unique sound and rhythm, depending on how the wind struck it. It was a 360-degree concert. The children heard it and it was as if a magical spell had been cast. They all stopped running in their tracks. I told the children to lie down where they were and listen. To my surprise, they all laid down for five minutes. They became still and completely entranced by the sounds. The wind stopped as if by magic and the sound event was over. The children popped up and created a new event called "catch us if you can."

NOTES

I play music with a band of avant-garde musicians and poets. During rehearsal, David Arner, one of the musicians, passed out a composition of his which consisted of three sentences of poetic music instructions. As he was passing it out, he said that there were notes. I was surprised that composers still used bar lines and notes. I was wondering what George and Chuck were going to do with the notes because they are not used to dealing with notation. When I looked at the page, I saw no musical notes. The page David passed out had only writing and I kept looking for the notes. I thought they might be invisible notes. I finally asked David, "Where are the notes?" He pointed to the bottom of the page where it said "Notes" and below that were two footnotes.

QUANTUM CHORDS

I was giving a concert at Indiana University with Franz Kamin, an avant-garde composer and friend of mine. We needed a 100-foot electrical extension cord for our concert. Franz said, "Go ask Dave. He has a lot of stuff in his office." Dave Baker was head of the jazz department and an exceptional trombonist. I ran into his office and blurted out that I needed a 100-foot cord. He was sitting at his desk and looked at me like I wasn't there. He was in a far-off place. After what seemed like a very long time, his eyes met mine and he said, "I don't think I've ever played one."

INDIANA TWANG

When I moved from Indiana to New York City, the tempo of my life drastically speeded up. I remember flying home after my first year in New York. My parents met me at the Indianapolis airport. I couldn't believe how slow my parents were moving and talking. They seemed to be talking with an accent. I thought something was wrong. I felt trapped in a record, playing at slow speed. I wanted to talk but I couldn't talk, and a pressure built up inside me. Then something "popped," and I felt myself shift into a slower Indiana rhythm. I began laughing at the sound and tempo of my voice. For the first time in my life I became aware of myself talking slowly with a "Hoosier twang" accent.

MASTER PIANO CLASS

I took a master piano class with Sophia Rosoff in New York City. When I was a child, I used to just play and make up stories. When I learned how to read music, everything changed. Improvising was not part of the classical approach to the piano. I went to Sophia because she was an exceptional classical pianist while also being open to improvisation, which is very rare in classical music. I improvised for her and she said, "I have nothing to teach you. Now play the first movement of a Beethoven Sonata." I would play for a few minutes and she would say, "Do you hear the difference between your improvisation and how you play Beethoven?"

Something about sitting with her made me aware of how hard I was trying when I played Beethoven, compared to when I was improvising. I wasn't playing Beethoven like I was when I was improvising. Even though I was playing from memory, I was still playing notes on a page. My rational mind was getting in the way of my musical self. Sophia did everything she could to get me to "just play."

She wasn't interested in "right notes," only in sound and what she called emotional rhythm. She would blindfold me, have me play every note wrong, and then just play the first note of each bar, which was called "outlining."

One day she didn't feel well and had a headache. I told her about my touch therapy skills and she asked if I would work on her. She laid on her couch and I worked on her head with a form of soft touch called cranial work. I naturally listened deeply into her system. After awhile she started to cry and her body went into a deep level of relaxation. She sat up and looked at me with the brightest eyes and said, "Just touch the piano the way you touched my head." From that point on, everything changed.

It was something I had been doing at the piano since I was a child.

I soon made the discovery that listening through my finger tips was much like listening with my ears. **When I played the piano as a child I would imagine something, listen to my imagination, and my fingers would naturally play what I was listening to. It didn't take long for me to perceive a major difference between a person and a piano. The difference is consciousness.**

When I played the piano, consciousness resided within myself and was expressed through the piano. Although different pianos had different actions and sounds, they didn't have consciousness. The people I touched like a piano had consciousness.

EINSTEIN'S VIOLIN

Composers, musicians, and poets make their living letting go of control, whereas scientists make their living holding on to control. During my time working in Switzerland, I sometimes imagined the nice Swiss couple who lived next door to Albert Einstein. They politely knocked on his door to complain about the sound of his violin and Einstein, with his unkempt hair, told them he was conducting thought experiments and riding waves of light into outer space. I wonder to this day if they would have reported him to a mental health professional.

SOUND HOARDING

I often joke in my classes that I am a sound hoarder, always in search of a higher power. Everyone likes to be in my studio and play with the sounds. I have hundreds of tuning forks on display, 20 crystal bowls, five crystal pyramids, 16 metal bowls, a Swiss sounding money bowl, seven Peruvian whistling vessels, four element chimes, as well as my 1914 Steinway grand piano, four Layne Redmond tambourines, and many flutes, drums, rain sticks, and whirlies. In the center of everything is my telepathic Dream Synthesizer with a large double terminated Brazilian quartz crystal. And if that is not enough, there are 16 drawers filled with noise makers, harmonicas, rattles, bells, esoteric sound instruments, and my collection of sheet music. I also have nine orgone crystal pyramids, numerous gemstone pyramids, and hundreds more crystals and gemstones placed in special geometric patterns around the room.

Sound and Intention Stories

SOUND AND SPACE

In 1997, I went for another master piano lesson with Sophia Rosoff in New York City. Her apartment was small and in the center was a Steinway concert grand piano which took up most of the living room. She asked me to play the first movement of Beethoven's *Apassionata Sonata*. I played the sections marked *forte fortissimo* as loud as the piano could produce. She stopped me and said, "Adjust your dynamics to fit the space." She had me stand up, close my eyes, and feel the space of her apartment. When I began playing again, my sound dynamics were congruent with the space. Every performance space is different, and a sound that is quiet in one space may be loud in another space.

HONKING HORNS

I came to New York City from Indiana University in 1973. It was a real shock. During my first months in New York I drove my Volkswagen bus around the city because I didn't understand subways. I got into honking my horn and projecting the horn sound to different people. The city was so filled with sound that it was often difficult to hear the horn. I would press the horn, project the sound, and watch peoples' responses without the feedback of the actual horn sound. I got very good at it. One day I took my car in for a safety inspection and they informed me that my horn was broken. I said, "What do you mean, I honk it all the time?!" The mechanic looked at me like I was crazy and said, "Broken, like it doesn't make sound!" I said, "How long has it been broken for?" He looked at the rusted wires and said, "For a long time."

DR. JOHN REEVES WHITE

In 1975, Franz Kamin and I were having dinner at my loft with Dr. John Reeves White. Dr. White was director of the New York Pro Musica Antiqua ensemble and a professor of musicology at Hunter College. I told Dr. White my "silent horn" story. He listened intently, smiled, and said that it made perfect sense because sound followed intention. Then he said, "Let me demonstrate." Dr. White sat at my piano and played a sound. Franz and I heard Dr. White alter a sound while it was moving through the air with his intention. He did this over and over, causing different effects. Franz suggested that he play a sound and Dr. White altered it with his intention after it left the piano. Dr. White caused the sound to "shimmer."

VLADIMIR MAXIMOV

I went to Brighton Beach, NY, in 1986 to visit the Russian sound healer Vladimir Maximov. Vladimir came from a remote part of Siberia where he learned ancient Russian energy techniques. In the 1970's he worked with many prominent Russian scientists investigating bioenergetic forces. Vladimir discovered that his healing energy could be layered onto the acoustic and video energy in audio and video tapes. For example, Vladimir would play a recording of the pianist Arthur Rubinstein and, while the recording was playing, he would embed the sound with his healing energy. Tuning into Vladimir and watching his body move while he was embedding healing intention was an amazing experience on many levels. In Russia, Vladimir was a sensation. He had a huge following of people who purchased his music recordings which had been embedded with his healing energy.

When trained musicians go back to the basics, they can discover sound healing. They must be willing to let go of their music training and become like children listening to sounds.

BEING LIKE A CHILD

I discovered the piano at the age of three. I lived with my grandparents and they had an old upright piano in their parlor. I remember sitting at the piano and exploring sounds. I went again and again to the piano. I couldn't keep away. I didn't know anything about music, notes, music theory, technique, or performing. After awhile the sounds became adventures and I would get lost in magical sonic journeys. The piano became my best friend.

When I worked at Bellevue Psychiatric in New York City, the hospital gift shop always had psychiatric patients' paintings on display and for sale. I got the idea that we should have patients' sounds in the gift shop. At that time I was working with a 25-year-old woman suffering from schizophrenia. I would sit at the piano and play for her and she would talk. One day I told her that she should play. She protested that she didn't know how to play. I instructed her to just make sounds like a child and not worry about making music. I told her to tune into her feelings and play sounds that were like her feelings. For a month she played her feelings and we talked. I started recording her and then convinced the gift shop to play the recordings in the background while people looked at the patients' paintings. In the next six-month period, we sold 34 of her cassettes.

PLAY LIKE A CHILD

When I was a child sitting at the piano, I didn't know how to control my hands or how one hand related to another. I sat at the piano and the sound moved them. Something inside of me—exciting, precious, and without a name—traveled within the sound and sought sonic relationships beyond consonances, dissonances, and virtuosic expectations. My hands, as extensions of my being, naturally and without effort, discovered sounds that inspired my imagination to soar. I navigated the sounds, discovering over and over again wonderful sensations embedded within waking dreams of magical realities.

IMPROVISATION

In the early 1980's I met with a group of new music composers and musicians at a performance space in downtown Manhattan called The Kitchen. The purpose of this group was to explore music improvisation. In the beginning, we would arrive at a set meeting time and play music. Our intention was to be with each other, listen deeply, and allow sounds to spontaneously emerge. We weren't interested in musical structures or in creating harmony and/or atonal music. We wanted to become vehicles through which sounds could freely emerge from within a communal space of listening. After many sessions, we realized that by having an agreed upon time to meet, we were imposing a structure on our improvisation. We all agreed to continue the improvisation group; however, we would not have any agreed upon times to meet. To this day the group continues to freely improvise.

SECTION TWO

Sound Healing and Values Visualization Process

The Sound Healing and Values Visualization Process is presented in four parts. Each part can be understood separately and can be read in any order. However, in practice, they merge, overlap, and integrate in many different and creative ways. The parts are:

PART 1: Visualization

PART 2: Values

PART 3: Feeling Tone

PART 4: Sound Healing

Prelude

Formless and unknown,

Receptive and listening,

Asking-giving-listening-asking-giving-listening,

There is a direction because I say so.

Listen, follow, and then forget what I said.

My saying is already unformed in the silence of your discovery.

Say for yourself and listen to the echo of your word.

When there is none, then you said yourself well.

There is no direction,

So make one.

If you don't like it,

then make another.

Let the adventure begin.

PART 1

Visualization

Go confidently in the direction of your dreams. Live the life you've imagined.

— Henry David Thoreau

Having goals, dreams, and visions to improve our lives and to enhance our wellness is a natural part of living. We vibrate with excitement when we envision reaching a goal or dream about something we want to have or to achieve in our life. The term "envision" means to inwardly see. It is our nature to continually envision new life possibilities. Envisioning is so natural that we use many terms to describe the different forms it takes:

- *Visualization:* I visualize myself having this in my future.
- *Imagination:* I imagine or have an image of this in my future.
- *Fantasy:* I have a fantasy of this in my future.
- *Dreams:* I am dreaming about having this in my future.
- *Projection:* I can see myself in the future having this.
- *Wishing:* I wish I had this in my future.
- *Wanting:* I want this in my future.
- *Pretend:* I am pretending that I have this in my future.
- *Desire:* I desire this in my future.
- *Make-believe:* I am make-believing that I am doing this in my future.
- *Needing:* I need this in my future.
- *Goal:* I set this goal for my future.

When we envision something, we see ourselves in the future living what we are envisioning. For example, we can *wish* for the sun to shine on a cold and rainy day which can lead to a *fantasy* of being in Florida on the beach. You say to your partner, "I *want* to go to Florida." Your partner gets excited and says, "Let's set a *goal* to be in Florida next year." You say, "That's great, I am really *dreaming* of being there. I wish I were there right now." She says, "We *need* to have this in our life."

Neurologically, the process of visualization begins as a thought of the future in the prefrontal cortex. The thought is projected onto the posterior cortex as sensory imagery.[1] We see, hear, smell, and taste ourselves living the thought.

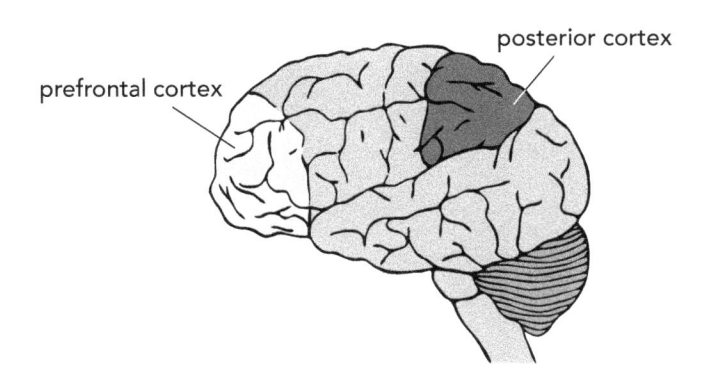

The prefrontal cortex is a part of the frontal lobe of the brain.

For example, a thought of owning a new car in the future is projected onto the posterior cortex as seeing, feeling, and hearing yourself driving your new car. If you change your thought to another car, the projection will automatically change to the future experience of driving a different car. The principle is simple: Every visualization creates a resonant projection. We are free to visualize the future we want. The process is much like going into a clothing store and trying on different shirts. We think, "I like that shirt." We put the shirt on and visualize ourselves in the future wearing the shirt. When we put on another shirt, our future visualization changes.

Although visualization is about an imaginary future, it has a very powerful effect in the present. When I was a teenager, the state of Illinois implemented one of the first state lottery systems in the Midwest. My parents would drive from Indianapolis, Indiana, to a small town just across the border in Illinois to purchase a lottery ticket. Teenagers usually don't want to go anywhere with their parents but I always wanted to go.

In those days, my parents were having financial challenges. There was always an unspoken tension in the house that sometimes erupted into arguments. Like magic, the moment we pulled out of our driveway to purchase a lottery ticket everything changed. The tensions of financial stress disappeared. During the whole trip, we talked about what we were going to do if we won. I could see my parents' facial expressions change and I could hear the sound of their voices become stronger and more resonant. It was exciting and fun. Even more fun was coming back and being able to hold the lottery ticket. It was like a sacred artifact to me. I touched it with the greatest care, making sure to protect it. The drive back was like floating on a cloud of infinite possibilities where anything was possible. Our car was like a magic carpet and I held the lamp that contained the genie in my hand.

Today I enjoy having an ice cream in a shop in upstate New York that sells lottery tickets. I sit and watch people before and after they purchase their tickets. When they come into the shop their shoulders are sagging and their faces are taut. After purchasing their lottery ticket they smile, stand straighter, and their faces relax and have more color due to increased blood supply. The odds are that they won't win the lottery but they keep coming back to buy tickets. It isn't surprising that dopamine reward pathways are thought to be very much a part of the addiction process. Lottery ticket buyers are not addicts, but they have found a high by visualizing what they'll purchase with the money they might win in the future.[2]

Neurologically, it appears that a positive visualization, like buying a lottery ticket, activates dopamine via the mesolimbic pathway, also known as the reward pathway.[3] Dopamine is a neurotransmitter that creates a heightened sense of calm and well-being. The reward pathway begins in the prefrontal cortex, travels to the nucleus accumbens, and then to the ventral tegmental area of the brain. The nucleus accumbens is sometimes referred to as the brain's pleasure center and has a lot to do with motivation and goal-directed behavior as well as addictive behavior. The ventral tegmental area serves as a dopamine production center that reinforces behavior.[4]

In non-anatomical terms, having a future visualization activates the reward pathways and creates a very powerful pleasure physiology in the present. In this sense, visualization can be used for healing or for getting high. It's important to recognize and to know the differences between values visualization and getting high visualization. The word "fantasy" comes from the Greek word *phantazein*, which means to make visible or to visualize. Fantasy is a huge business. Fantasy sports are estimated to have 32 million worldwide followers who spend three to four billion dollars a year. Perhaps the greatest fantasy of all is the lottery. In 2012, Americans spent $65.5 billion dollars on lottery tickets. Although we like to believe that winning the lottery is obtainable, the odds of winning are 1 in 175,223,510.[5] If a fantasy is unobtainable, then why do we continue to fantasize? The short answer is that the process of fantasizing gives us a dopamine reward that is more important then what we are fantasizing about. Values visualization integrated with sound healing also gives us a dopamine reward. The difference is that the dopamine reward is part of a larger conscious value system that is associated with obtainable goals.

VISUALIZATION EXERCISE

1. Become aware of and describe as accurately as possible what you are visualizing. For example:

   Wants: I want a new car, a new house, a new pet. I want to go to school and receive a diploma. I want to win the lottery for twenty million dollars. I want to go on a trip around the world. I want to win a sports tournament. I want to play an instrument.

   Goals: My goal is to eat less sugar. My goal is to wake up every day and do yoga. My goal is to enroll 20 people in my new class. My goal is to swim a mile a day.

   Fantasy: I see myself as being as light as a feather and floating in the clouds. I am the king of a magical kingdom and everyone comes to me for advice. I am like superman—I can fly and see through walls and no one can hurt me.

   Needs: I need for people at work to appreciate me more. I need for my children to go to bed on time. I need to take a break from the computer every hour. I need to travel more.

2. Once you have stated *what* you want in these terms, then you can learn to use sensory-based terms, i.e., seeing, feeling, hearing, tasting, and smelling, to describe your vision. Tune into your vision and list two to five sensory words that are congruent with your vision.

   For example, if you visualize being on a new boat sailing the Caribbean with friends, describe this.

   Touch: Feel the boat rocking on the ocean waves. Feel the wind on your skin.

   Sound: Hear the sounds of the ocean or your family talking.

   Sight: See the sun or moon reflecting over the ocean. See the stars.

   Taste: Taste the food you are having in a good meal with family and friends.

   Smells: Smell the sea air. Smell the scent of a dinner cooking.

3. Once you have described your vision, write it down in sensory terms.

4. Describe your vision to a partner using your sensory terms as though you are experiencing it right now.

5. Close your eyes and have your partner repeat your written vision back to you in your own sensory terms.

	Sight	Sound	Touch	Taste	Smell
1					
2					
3					
4					
5					

PART 2
Values

Once you have tuned into your visualization, the next step is to slow down and determine the values your visualization expresses. A value is defined as what is most important to you in your life. The story of the lottery is a story of values. Lottery winners who are not prepared for winning rarely have a defined value system and community support network. Seventy plus percent of all lottery winners squander away their winnings in a few years, resulting in the destruction of family and friendships.[1] Fortunately today, many states suggest and provide opportunities for large-sum lottery winners to receive emotional and financial counseling to prepare them for the shift in their lives.

A fundamental part of their counseling process is to help them become aware of their financial resources and to be able to use those resources as an expression of their life values. Money without life value is empty of meaning. If we know our values, money can enhance their expression. For example, a person who values teaching was teaching a class on wellness at an adult education program at his local school before winning the lottery. After winning the lottery, he built a wellness teaching center, hired a staff, and created a wellness educational program and clinic for his community. He continued teaching and expressing the values that were most important to him. Values clarification is a process designed to identify the values contained in a visualization.[2] The following chart shows the difference between knowing and not knowing your life values.

Value-Deficit Behaviors	Value-Driven Behaviors
Apathy	Purposefulness, enthusiasm
Flightiness	Commitment
Inconsistency	Consistency
Over conforming	Critical thinking
Over dissenting	Balance of conformity and individuality
Unsatisfying decisions	Satisfying decisions

Determining and Establishing Your Values

A value is something that you care deeply about and is very important to you. All visions, whether you think they are negative or positive, contain important life values. However, many of your values may be unconscious and based on beliefs passed on to you by your parents, friends, and authority figures. The values clarification process brings these values to consciousness in order to ask yourself if they are truly your values. At the same time, the values clarification process empowers you to create new values or to modify old values to be congruent with what is most important to you in your life.

To begin, you will need a pencil and paper.

1. Sit quietly by yourself or with a person or persons you trust. The values contained in your visualization must be chosen freely and without peer pressure. They must not be coerced and they must be meaningful to you. Therefore, it is very important in this step to feel that you are safe. Any fear you have will reduce the likelihood of your values being chosen freely.

2. Have fun. Share your vision and brainstorm potential values that your vision is expressing. You can create many values that your vision may be expressing. Don't put pressure on yourself to create the "right values." Create a list of these values. Writing down your values as they come should be a free process. It's not necessary to make a formal list. Just discover a value and write it down, scribble it, make it larger or smaller, etc. If you are working with someone you trust, they can help you brainstorm your values. Instead of writing them down, you can speak them into your smart phone recorder.

3. Look over your values and make a more organized list. As you look at your list more closely, it's OK to use your critical mind to evaluate each value in terms of importance. Ask yourself if your values are congruent with your life goals and if they are supportive of those around you, including family, friends, and community. Evaluate the pros and cons of each value and its consequences. Be free to change and modify your values as you go through the process.

VALUE IDEAS

Accountability	Environmental	Making a difference
Achievement	Efficiency	Mentoring
Adaptability	Ethics	Open communication
Ambition	Excellence	Openness
Attitude	Fairness	Patience
Awareness	Family	Perseverance
Balance (home/work)	Financial stability	Personal fulfillment
Being the best	Forgiveness	Personal growth
Caring	Friendships	Power
Coaching	Future generations	Professional growth
Commitment	Generosity	Recognition
Community	Health	Reliability
Compassion	Honesty	Respect
Competence	Humility	Responsibility
Conflict resolution	Humor/fun	Risk-taking
Continuous learning	Independence	Safety
Cooperation	Initiation	Self-discipline
Courage	Integrity	Success
Creativity	Intuition	Teamwork
Dialogue	Involvement	Trust
Ease with uncertainty	Job security	Vision
Enthusiasm	Leadership	Wealth
Entrepreneurial	Listening	Well-being

4. From your values list, rank the values in order of importance. For example, a client of mine who was using her living room as a studio said that if she won the lottery she would buy a studio with lots of space. I asked her what she would do in her new spacious studio. She said (the values she later determined are in parentheses) that she would have special areas where her different creative projects could take form (*creativity*). She would be able to move around freely from project to project (*space*). Wherever she was working, there would be a feeling of "mine" and the whole space would reflect herself in a special way (*warmth*). She further said that she would invite friends into her space and could see herself having a good time sharing and being together (*sharing/community*). She concluded by saying that the whole space would be a fun space where new ideas would always be welcome (*fun*).

When she was asked to determine the values expressed in her vision of a bigger studio space, she ranked them in the following order. Remember that you can always change your rankings. If one value is close to another value, do your best to put them in order. Or you can know your top three or four values which are the most important regardless of how they are ranked.

1. Creativity
2. Space
3. Warmth
4. Sharing/community
5. Fun with friends

5. In order for the values within your visualization to become real, they must be consistently acted upon. We must demonstrate our values through our behavior. Up until now we have been talking about and clarifying our values. In this step, the values you arrived at must be valuable enough to live via your daily actions. In other words, you must "talk the talk and walk the walk." A feedback loop takes place when you act on a value and you consciously know that you have acted on that value. The feedback loop activates the reward pathway and creates a cascade of biochemical neurotransmitters, including dopamine, that enhance your health and well-being. However, acting on your values is not a guarantee of a perfect life without challenges. It's always a risk to express your values and get feedback from those around you. Hurt and emotional challenges are much easier to learn from in the context of conscious value expression.

Creating Value in Different Ways

It's important to understand that our values can be expressed in many ways. It's important to affirm to yourself: *I am open to my values appearing in many different forms and expressions.* This requires tuning into our values and asking ourselves if we are already expressing and acting on them in ways we may have taken for granted and/or have not been aware of. For example:

- I invite my family to my home for Thanksgiving dinner every year. (*family*)
- I enjoy making my house ready for everyone to come visit and I enjoy baking the turkey. (*generosity*)

- I go bicycling every Saturday with my friends. (*friendship*)
- We take different trails and there is always something new to see. (*adventure and appreciation of nature*)

Once you have determined your values, look over your list and tune into ways you may already have been acting on them. Sometimes we "get the cart before the horse." The horse represents our values. The cart represents the form we are expressing them through. The cart before the horse is waiting for a specific form to express our values. For example, if only I had a studio I could paint. The real value is what comes from painting. The studio and paint are the cart. There are many carts. The following case story illustrates this.

NURSING HOME CASE EXAMPLE

I was conducting reality orientation therapy with a group of older patients at a nursing home, assisted by a staff nurse. I asked my patients to share what they wanted most in life. A man in his late 80s said that what he wanted the most was to be a fullback for the New York Giants. I asked him what it would be like if he were. He said he'd be excited, run full speed ahead, and that everyone would cheer for him. His eyes lit up as he shared this. The nurse became upset and asked to talk with me in another room. She said that I should be careful because he could become depressed. I asked her why and she said that it was impossible for him to be a fullback. This wasn't reality.

I agreed with her that it wasn't possible for him to be a fullback for the New York Giants. I also pointed out how animated and excited he was, and that this was a goal of reality orientation. Therefore, it was important that we create a way to make it real for him. We went back into the room with the patients and I again asked the man what it would like to be a fullback for the New York Giants. He said, "I'd be free and no one could tell me what to do."

The patient was expressing his value of freedom and the importance of free expression in his life.

I said, "If you were a fullback, you'd still have to know the plays and your coach would tell you how to run them."

He said, "That's OK because when I run I'd just go all out."

I then said, "If you had the freedom to go all out, would you care whether you were a fullback or not?"

It took less than a second for him to say, "No."

Sometimes we have to modify or give up the initial form of a vision in order to achieve the essence of the vision. In my patient's case, the initial form of his vision, being a fullback for the New York Giants, was clearly not achievable. However, his vision was important because it expressed his important personal values of being strong, focused, and receiving acknowledgement from others.

The next day I got a set of plastic bowling balls from the recreation room and set up a special bowling alley. I seated the patients in his reality orientation group, plus staff and visitors, on both sides of the bowling alley. I gave my patient a plastic bowling ball. I told him I was his coach, and that he should roll the ball and knock over the pins. However, before he could roll the ball, he should make eye contact with the group and show them his focus.

He was into it and the group, including the staff, clapped and cheered. He looked at the bowling pins with focus and rolled the bowling ball. The ball moved slowly. I thought it would never reach the pins. To my surprise and to everyone's amazement, he knocked over every pin in slow motion. We all cheered! To this day, I cannot figure out how all the pins fell over. We talked about it for weeks in our reality orientation group.

I designed an experience that had the potential to manifest his values in a different form. Knowing our values is a way of checking the validity of an experience. There are many love stories built around value testing, where the man or woman has a specific vision of the person, "the right person," to fall in love with. They find the person that looks like their vision, but the relationship doesn't work. Next, they meet someone, usually by chance, who seems totally different than the person they had visualized. Over time and through a set of awkward encounters, they both come to realize that they share common values and are truly meant for each other.

PART 3
Feeling Tones

Our visualizations create very real vibrations in the present; we call them "feeling tones." Feeling tones are the vibrational component of visualizations that originate in the limbic system of our brain and travel throughout our body as waves of vibration. When we learn to tune into feeling tones, we can start to feel an emotional tone moving throughout our body. To better understand this, press an electric massage vibrator to your body and feel the vibrations. Take a deep breath, relax, and just be with the vibrations without any emotional associations. The vibrations moving through your body are pure wave sensations. Surrendering to the wave sensations will bring your consciousness into a dimension beyond emotion, meaning, history, theory, belief, or imagination. The vibrations from a vibrator originate outside of your body, whereas the emotional vibrations of a visualization originate within the limbic system of your brain.

The following diagram is based on the joints of the body, which can be imagined as nodal points. Waves of feeling tones travel through these nodal points, and each joint must be flexible in order to conduct different feeling tone waves.

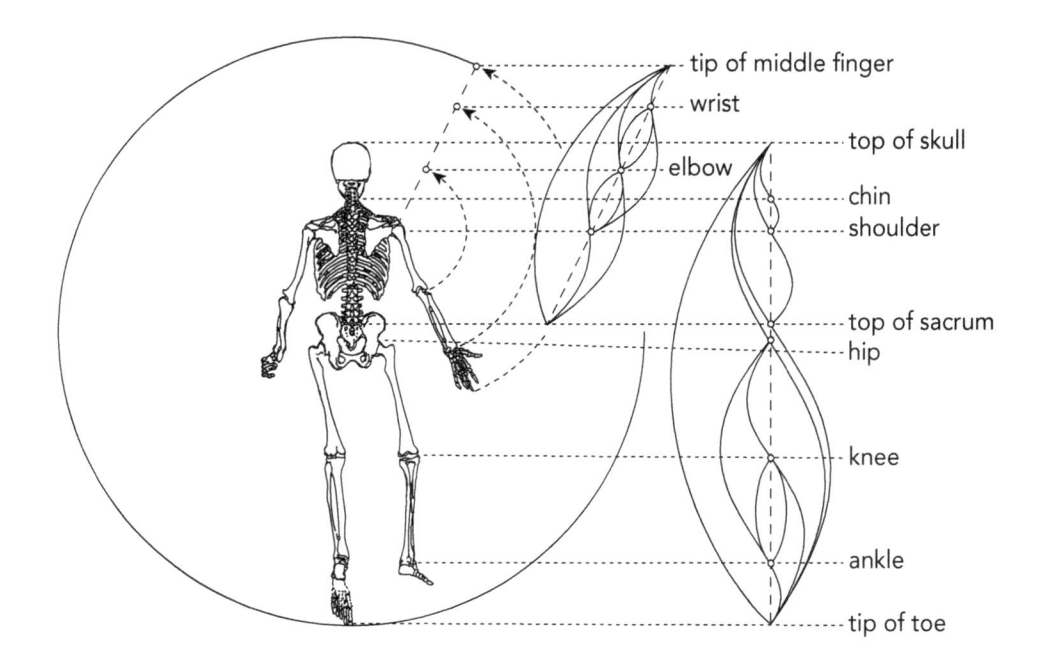

An inability to conduct feeling tone waves creates tension in a joint or joints which can lead to mental, emotional, and physical dissonance. For example, somatic psychotherapists help us resolve emotional challenges by asking questions like, "What sensations are you feeling in your body right now?" These somatic questions are designed to help a person disassociate from emotional memories and to become aware of feeling tone sensations. Dr. Wilhelm Reich, a pioneer in body psychotherapy, used the term "streaming" to describe the sensations of emotional waves of energy moving freely throughout the body during a therapy session.

Dr. William Grey, a psychiatrist, created the term "feeling tone" to describe the sensations of emotions moving through our body.[1] When we say "I feel sad" or "I am happy," we are describing sensations moving through our body that we have associated with emotional experiences. If we sit quietly and recall a time when we have experienced different emotions, it's possible to measure the feeling tone waves as signatures.

The Sentics research method, developed by Dr. Manford Clynes, measures finger movements related to emotions.[2] Subjects place their middle finger on a Sentograph which independently measures vertical and horizontal components of finger pressure through specially designed transducers. The subjects are asked to visualize different life events and recall the emotion they felt. The results are automatically graphed in a computer and correlated with other results.

Finger placed on Sentograph

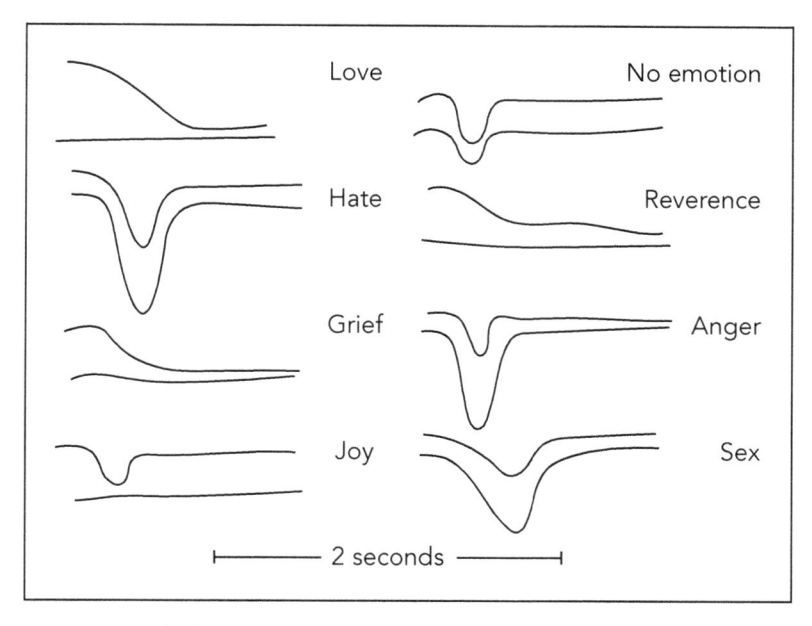

A Sentic graph showing emotions as waves that move through our body.

Understanding an emotion as a feeling tone and then trying to get rid of it is like trying to get rid of a guitar string and then expecting to play good music. We need all of our emotions to play the music of life. All of the emotional responses listed below are also feeling tones. Emotions and, more importantly, feeling tones are normal and necessary parts of the healthy vibrational flow of energy through our mind and body. Healthy emotional responses that flow through our body as feeling tones are like currents of flowing water. Naturally flowing water is pure, clear, spiraling, and fundamental to the quality of life. Interrupting and manipulating the natural flow of water creates stagnation which can become a breeding ground for bacteria and parasites. Here are some emotions that create feeling tones:

Anger	Hate	Shame	Guilt
Worry	Regret	Surprise	Frustration
Fear	Loneliness	Contempt	Awe
Sadness	Exhilaration	Distress	Relief
Happiness	Dejection	Grief	Empathy
Despair	Envy	Joy	Hope
Embarrassment	Resignation	Thankfulness	Pity

In Ayurveda, the natural medicine system of India, emotions are called *rasa*. Rasa is a Sanskrit word which describes the energy of human emotion as a juice that colors the mind. It is congruent with the term "feeling tone." Ayurvedic physicians describe nine basic rasas, all of which must be kept in balance for optimal health and well-being. Ideally, a rasa moves through the body as a wave of vibration. Indian musicians play special ragas at different times of the day to balance the rasas. The rasas and their Western emotions are:

- Sringara: love, unity, beauty
- Hasya: humor, joy
- Adbhuta: inquisitiveness, wonder
- Shanti: calmness, peace
- Raudra: anger
- Vira: confidence, power
- Karuna: sorrow, empathy
- Bhayanaka: apprehension, anxiety
- Vibhatsa: self-pity, disgust

In Christianity, the transformation of emotion into feeling tones of value is called "the passions" or "the passions of Christ." Christ knew his mission and what he was called to do. He said in Luke 14:28: "Which of you, wishing to build a tower, does not first sit down and count the cost to see if he has the resources to complete it?" Building a tower means to act on our vision, move towards our goals, and accomplish our dreams. To do this, we must sit down and count the cost. This means we must take the time to reflect on our vision and consciously determine if we are willing to pay the cost to achieve our value. It is an ancient cost-to-income ratio analysis: What will it cost me to build my tower and what value will it bring me?

We enjoy feeling tone orchestrations all the time. Children, when they sit on Santa's lap and share their Christmas wish lists, are creating a concert of feeling tones that can be observed as waves of vibrations moving through their body. Each gift they wish for creates a new wave. Children also love to go fishing and catch fish. When the fish bites and pulls on the string, they scream in excitement. When they pull the fish from the water they become so excited that they jump up and down. When they learn to identify different fish they create different fish feeling tone orchestrations. For example, they may have bass excitement, catfish excitement, or trout excitement. Each excited state is a different feeling tone frequency.

I was fishing in Florida with my twin boys. One of them caught a puffer fish. They had never seen a puffer fish and didn't have a name for it. When they lifted it out of the water, it puffed up like a balloon and sailed away on the wind then back into the water where it deflated and disappeared. To this day if I say "puffer fish," they immediately become like tuning forks, vibrating the feeling tone frequency of their puffer fish excitement. If I said, catfish, they would be excited; however, it would a different feeling tone frequency than puffer fish.

From a sonic perspective, feeling tones are organized vibrational patterns that have the potential to reorganize our mind and body. The science of cymatics was developed by Dr. Hans Jenny to demonstrate the relationship between geometric patterns and vibration.[3] Dr. Jenny performed cymatic experiments by putting substances such as sand, fluid, and powder on a metal plate attached to an oscillator where the vibratory rate is controlled by a frequency generator. During his experiments, Dr. Jenny could hear the sound and simultaneously see it create geometric forms on the vibrating plate. By changing the frequency of the underlying vibratory field in the plate, he was able to observe, in real time, and effectively demonstrate the organizing design nature of vibration.

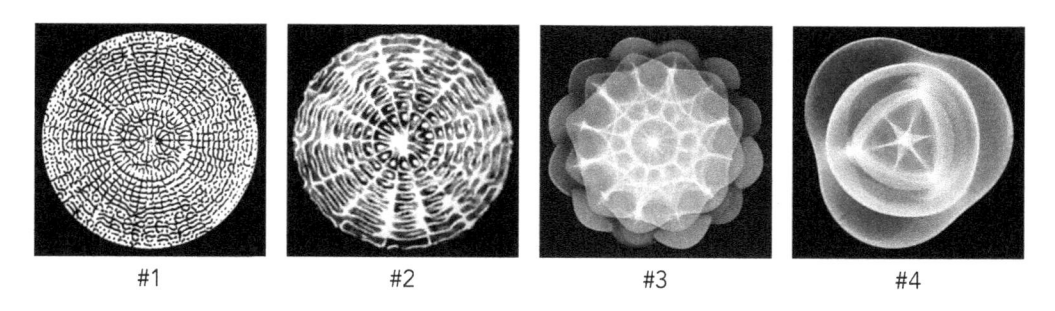

#1   #2   #3   #4

These images began as a pile of sand (#1 and #2) or a drop of water (#3 and #4) on a metal plate. When the plate was vibrated to an audible frequency, the sand or water consistently moved into the same geometric pattern based on the frequency. Although the photos appear to have been drawn, the geometric patterns are all created by vibration. If one were to attempt to change or rearrange the geometric pattern without changing the vibrational frequency, the original geometric pattern would always return.

We can use cymatic imagery to understand the effect of visualization on our body. Every time we visualize something, we create a feeling tone that moves through our body like a wave. Much like the sand or water organized by vibration in a cymatics experiment, feeling tones geometrically organize our body to match the geometric pattern of the feeling tone. In other words, our bodies geometrically form the shape of the vibratory pattern of a feeling tone. We can observe this in children's bodies and, if we look, we can see it in adult bodies, as well as learn to feel it in our own bodies.

The alignment of our body to specific feeling tones has tremendous healing potential. Chiropractic medicine is based on aligning our body to geometric patterns based on tone. Dr. D. D. Palmer, the founder of chiropractic medicine, said, "Life is an expression of tone and that consciousness is determined by and must accord with acoustic vibrations."[4] Although greatly misunderstood at the time, today his understandings of tone and the body are congruent with modern quantum research. Although it is a scientific given that we are vibrational beings in a quantum universe, the medical model has yet to accept this. At the same time, it is interesting to note that there has been a lot of research done that uses vibratory terms to describe how our body functions. The following research studies are only a few examples of what is being done that involve the body and vibration. The following studies are some of many examples.

- "Human brain networks function in connectome-specific harmonic waves"[5]
- "High Gamma Power Is Phase-Locked to Theta Oscillations in Human Neocortex"[6]
- "Theta Oscillations in the Hippocampus"[7]
- "Quantum cognition: a new theoretical approach to psychology"[8]
- "Melody discrimination and protein fold classification"[9]
- "Neuroplasticity Beyond Sounds: Neural Adaptations Following Long-Term Musical Aesthetic Experiences"[10]

In systems science, a feeling tone can behave much like a strange attractor. A strange attractor is a new vibrational pattern that has the potential to reorganize a whole system. A strange attractor is strange because it is a new pattern that is alien to the old system. The feeling tone of a vision is a strange attractor because it is an organized vibrational pattern that has the potential to reorganize our life, based on achieving our visualization. Every time we act on the values of our visualization, we are acting on a new vibrational pattern in our life. When the old system reorganizes into a new system, a feeling tone strange attractor is no longer strange. Intuitively, this is easy to understand because it happens to us all the time. For example, imagine you are driving through a new neighborhood and see a house you really want. Imagine yourself being in the house and how your life would be different. Living in your imaginary house is fun and you start going there in your imagination. Whether you are aware of it or not, you are activating a feeling tone strange attractor every time you visualize your new house. You may or may not get the house you are visualizing, but you always have the potential to receive the benefit of the values inspired by the vision of the new house.

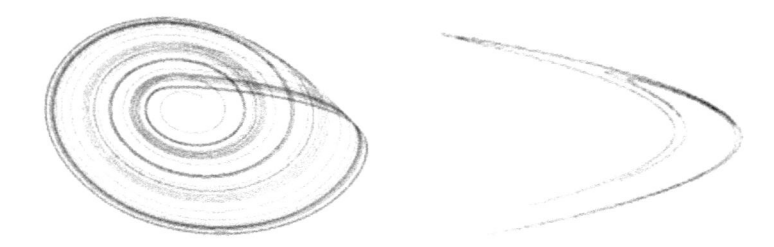

Computer images of strange attractors. We cannot see feeling tones anymore than we can see the geometric organizing patterns of a sound. However, mathematicians can model the visual appearance of strange attractors on computers.[11]

To put it simply: We are vibrational beings living in a vibrational universe. We are continually creating and conducting vibrations. Values visualization integrated with sound healing is a way of navigating our vibrational universe. We visualize to know where we are going. Visualization gives us direction and focus. We all want to go to a place of value. The term "value" can also mean "heart" and visualization can mean "path" as in "I see where I am going." The Yaqui shaman Don Juan eloquently and poetically summarized the importance of a life path with heart or what is called in modern research "value driven behaviors."

All paths are the same: they lead nowhere. However, a path without a heart (without value) is never enjoyable. On the other hand, a path with a heart (value) is easy—it does not make a warrior work at liking it; it makes for a joyful journey; as long as a man follows it, he is one with it. Look at every path closely and deliberately, then ask yourselves this crucial question: Does this path have a heart (value)? If it does, then the path is good. If it doesn't, it is of no use.

— Carlos Castaneda, *Tales of Power*

The idea of a path or visualization leading to nowhere at first seems nihilistic. What is the point if we are always going nowhere? However, upon closer examination, a path (visualization) is a vehicle and a container for expressing one's heart and values. If a path is no longer available to you, another path will come, and your heart will awaken and express value through the new path.

Archetype of the Two Lucifers

In the Bible, creation begins with the Word, and God next says, "Let there be Light." Lucifer is God's most beloved archangel and the bringer of light. The same Lucifer is also a fallen archangel and the devil. When we visualize (light), we invoke the archetype of the two Lucifers. Are we invoking God's most beloved archangel who is harmonious with "The Word"? Or are we invoking the fallen Lucifer who, as Milton describes in *Paradise Lost*, is the Lucifer who would rather reign in hell then serve in heaven?

When we use our mind to visualize, we illuminate what we want. For example,

"I want that new car."

"I visualize a house overlooking the ocean."

"I see myself in the future being married to a beautiful person."

Ideally, our visualization is a catalyst that can guide us into the deeper value and feeling tone of the vision. If we focus only on what we have visualized and do not clarify its value, we can get lost. The fallen Lucifer wants the vision exactly as visualized and will do anything to get it. The beloved Lucifer uses the visualization as a catalyst and seeks its real value in many different forms.

The fallen Lucifer is eloquently described by Black Elk, the spiritual leader of the Oglala Lakota Sioux, when he says, "It is in the darkness of their eyes that men get lost." We get lost in our visions all the time. I worked with a client who told me he was looking for a relationship. I asked him to visualize the woman he was looking for. His eyes popped open and he said, "That's easy." He opened his briefcase and showed me a picture of a beautiful model and said, "This is my vision of the ideal woman." I asked him if he had ever had a relationship with a woman who resembled his vision.

He said "Yes. I've been with several women that I thought were perfect".

I asked him, "What happened?"

He said, "I always found something that was not perfect about them."

My client was "lost in the darkness of his eyes." He wanted reality to conform to the external form of his vision. In extreme cases, a person might even stalk the model with only the photo, in hopes of manifesting their exact vision.

There are many cultural traditions in which a visualization process is used to enlighten oneself about the direction of one's life. The term "vision quest" was first used by 19th century anthropologists to describe the visualization process used by Native American cultures.[12] Although greatly misunderstood by anthropologists at the time, a vision quest is a community-supported planned process in which one systematically prepares for and enters into a visionary state for the purpose of gaining life guidance and value. In this context, the experience of visualization and feeling tone can be very powerful. In the book *Black Elk Speaks*, Black Elk describes his experience of feeling tone orchestration after a vision quest as a glow. He says,

> *…as I lay there thinking of my vision, I could see it all again and feel the meaning within a part of me like a strange power glowing in my body; but when the part of me that talks would try to make words for the meaning, it would be like fog and get away from me.*[13]

Black Elk can see everything; however, his focus is on the strange power glowing in his body. This is the feeling tone orchestration of his vision. Tuning into it is like listening to a beautiful piece of music and trying to describe it. You know the experience, but words are never adequate.

Feeling Tone Exercise

We can learn to tune into the feeling tone orchestrations of what we are envisioning through mindful listening and body awareness. Just as we navigate a visual universe with our eyes, we can learn to navigate a vibrational universe with our sense of felt body feeling. Through this increased awareness, we will be able to make better life choices about the effects of what we are envisioning based on immediate vibrational feedback.

1. Close your eyes, take a deep breath, relax, and imagine something you really want, for example a new house, a puppy, a new car, a vacation trip to an exotic place, etc. For this exercise, we will use a new car. Just substitute whatever you want for a new car.

2. Tune in, take a deep breath, and become aware of your body. Get an overall sense of the felt feeling of your body.

3. Imagine the new car you want. The sky is the limit because it is your imagination. For example: "I want a new Tesla Model S." Take a deep breath, close your eyes, and imagine driving your new Tesla Model S.
   a. See yourself looking at the car, getting into it, turning it on, and feeling the acceleration as you drive down the road.
   b. Say to yourself or your partner: "I have Tesla Excitement."
   c. Close your eyes and bring your awareness to the felt feeling in your body of Tesla excitement.
   d. Compare the felt feeling with the felt feeling of your body before you imagined a new Tesla.
   e. To further understand feeling tones as a vibrational orchestration, imagine different cars and feel the different vibrations of excitement in your body.
      Tesla Excitement
      Chevrolet Excitement
      Ford Excitement
      Lexus Excitement
      BMW Excitement
      Volkswagen Excitement
4. Describe your body awareness with each car either in your journal or to your partner.
5. Tune back into your felt body feeling of each car and hum a sound that resonates with your feeling. Don't hum a tune or try to make music. Just hum a pitch from high to low that is like the feeling tone of each car.
6. What music recording would you play while driving your Tesla S that would be congruent with your feeling tone?

Interlude

STARFISH DREAMS

The dream lives of starfish enter into resonate multi-dimensional oscillations,

    intertwining themselves into patterns of realized thoughts

Nearly silent sounds emerge from holes defined within the space of their

    interdimensional network.

Gaseous molecules drift into intervallic spaces between their star points.

Adorned, jeweled, awake and knowing, a volatile spark emerges igniting the

    intervallic gases into bursts of illuminated logical thoughts.

Being androgynous they listen with tantric ears

    embracing,

        enfolding,

            and dreaming.

PART 4
Sound and Values Visualization

The Values Visualization Process is designed to create a highly-tuned mental event focused on healing and the attainment of our life's goals. **When a focused intention of value is carried by sound into the core brain, the brain instantly responds and creates neural networks that are resonate with the feeling tone of the intention. Sound is used to create sonic fields in which listeners are free to discover, enter into, and move through continually forming and reforming geometric harmonic networks that mimic the geometric patterns of neural networks.** When the listener enters into the sonic field with a clear intention, these sounds act as carrier waves that transport feeling tones past the rational mind, past emotional and physical defense mechanisms, and directly into the core brain. There are two broad categories of sonic fields. These are ambient sonic fields which are created by natural or instrumental sounds without a persistent beat, and five element sonic fields which are created by natural or instrumental sounds with a consistent beat. The two categories of sonic fields can be discussed individually but, in practice, they can be combined and integrated in many creative ways.

Ambient Sonic Fields

<u>Naturally occurring ambient sonic fields</u> are created by sounds of nature such as ocean waves, flowing streams, waterfalls, rainfall, birds, frogs, crickets, and wind. Nature's sonic fields are primal sound baths that have been used for healing for thousands of years. When we go to a park or take a walk on the beach, we naturally appreciate these sounds. Today, we have access to high quality nature recordings. It is easy to go onto YouTube and search for your favorite nature sounds. Here are some nature sound experiences that students have shared with me.

Tree Frogs and Crickets

"I was looking for an answer to a problem I had been grappling with for some time. My thoughts were focused on what to do and my feelings were that I would never figure it out. I was feeling hopeless. It was a warm summer night and I couldn't sleep. I sat on my porch thinking more about my problems. I remember taking a deep breath and all of a sudden I found myself absorbed in the sounds of crickets and tree frogs. Suddenly I became aware of thousands of sounds all around me. I gave myself to the sounds and, as I listened, my problem was not so important. It was as though the tree frogs and crickets became intelligent voices creating webs of sound. I lost track of time, and when I returned I was refreshed and excited. Later that day, a friend called and said something that was totally unrelated to the problem I was working on and this led to a whole new way of dealing with the problem."

Here is another sonic experience that one of my students shared with me.

My Waterfall Rock

"I was sitting near a waterfall talking with my friends. Suddenly, something within the sound of the waterfall caught my ears. I tried to ignore it because I wanted to talk with my friends. The sound kept coming back. Finally, I excused myself and sat on a rock very near the waterfall. The rock felt like 'my rock' and sitting on it gave me an immense sense of security. I just let go and dissolved into the sounds of the flowing water. I kept discovering new sounds, and all of the sounds were somehow related to each other in the most intimate ways. I became a sound flowing with the other sounds. I realized that there were trillions upon trillions of sounds and that we were all flowing and communicating as one."

Instrumental ambient sonic fields are created by instruments without a persistent beat using tuning forks, gongs, crystal bowls, metal bowls, and bells. BioSonic tuning forks are designed to create an ambient sonic field consisting of thousands of overtones which are continually creating geometric harmonic patterns that resemble multi-dimensional neural networks. A study, published in *Frontiers in Computational Neuroscience,* uses topology to identify high-dimensional cavities (holes) that form when the brain is processing information. Neurons form geometric networks called cliques that are filled with cavities or holes that lead to different dimensional information processing. When the brain is stimulated by a sonic field of tuning forks, it reacts by organizing complex neural networks of multi-dimensional geometries to process the sounds.

The following drawings are of neurons in which the black dots represent the geometric patterns that are below each drawing. The formations of these networks are responses to different mental events, such as listening to sounds or introducing an intention embedded in a feeling tone. Neural networks can form and disappear almost instantly. Researchers describe the representational drawings as: "The progression of activity through the brain that resembles a multi-dimensional sandcastle that materializes out of the sand and then disintegrates."[1]

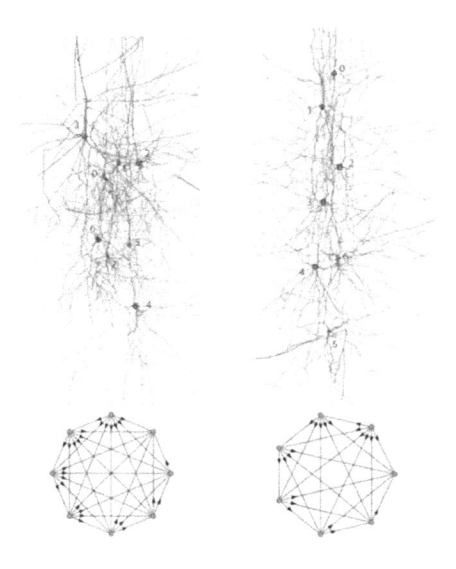

Below (*left*) is an oscilloscope picture of a sonic field which was created by tapping C & G BioSonic tuning forks (*right*). Within the sonic field, waves are continually moving and pulsing in different patterns that mimic the formation of geometric neural networks.

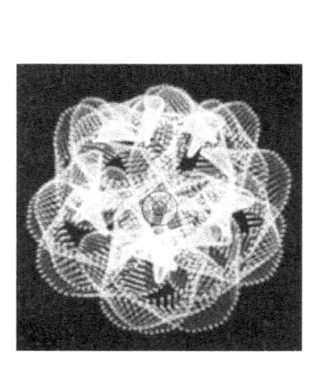

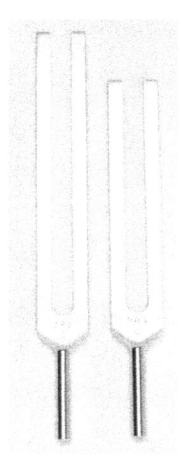

When the listener enters into an ambient sonic field created by using tuning forks, gongs, or sounding bowls, they can discover and entrain with continually forming and reforming harmonic networks of sound and their overtones. **When the listener enters into an ambient sonic field with a clear intention that is condensed into a feeling tone, their consciousness will be drawn to the sounds that resonate with the feeling tone of their vision.** These sounds will network and create a geometric pattern that resonates with the sonic geometric patterns of their feeling tone. The feeling tone pattern is carried via the auditory thalamus into the core brain where it is amplified into a whole brain resonance.

The human brain is a complex, multi-dimensional labyrinth of networks and passages in constant flux, based on a continual input of mental events. New neural connections are instantly and continually being created, pruned, and then deconstructed. Every day, billions of neurons vibrate and communicate with each other in harmonic waves traveling at light speeds through quadrillions of synapses. In 1996, Stuart Hameroff and Roger Penrose theorized a model of consciousness they called Orchestrated Reduction (OR) in which the brain was seen as a quantum computer.[2] Nearly 20 years later, the brain went from being a computer to a quantum vibrational orchestra. In 2015, Stewart Hameroff gave a talk for the Scientific Association for the Study of Time in Physics and Cosmology in which he said,

> "The brain is looking more like an orchestra, a multi-scalar vibrational resonance system, than a computer. Brain information patterns repeat over spatiotemporal scales in fractal-like, nested hierarchies of neuronal networks, with resonances and interference beats."[3]

From the "brain orchestra" perspective, a feeling tone is a "brain information pattern" that creates geometric neural networks. Within these networks our consciousness has access via nodes. These nodes are aligned with holes in the neural network, and are resonant with the harmonic information patterns in all dimensions.

When the listener's consciousness merges to a harmonic network congruent with a feeling tone, the geometric pattern of the feeling tone acts much like a brain algorithm. Algorithms are sets of rules that are followed in order to acquire a goal. For example, computer algorithms scan vast amounts of digital data in order to bring to our awareness information that is congruent with the rules of our search. There is so much data available on the internet that it is estimated it would take three million years to download it at current speeds. It is amazing that we can type in a few words into Google and data instantly "pops up" that is congruent with our search terms. We can change the search words, compare search results, and, in general, be neutral in our evaluation of data relating to our search.

When we initiate a values visualization search, we are searching an infinite field that includes our whole being and our quantum connections to the universe. The data that "pops up" in three dimensions is vibrational, may have originated from multi-dimensional sources, and is subject to our three-dimensional interpretations. In order to cognize and evaluate the results of a search in real life, we need to creatively interpret and organize multi-dimensional data, which may appear to have no resemblance to our vision, into a recognizable form that is congruent with our vision. This is illustrated in topology by the relationship of a donut to a teacup.[4] Different geometric forms are defined in topology by holes, just as different geometric forms created by sounds are defined by nodes. For example, a donut has one hole and a tea cup with a handle has one hole. Normally, we think of a tea cup as being different from a donut. However, if the donut were made of putty we could stretch, crumble, and bend it into a teacup.

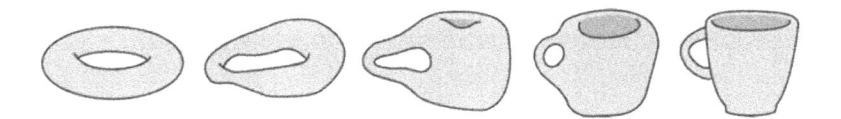

In other words, objects can take on many shapes and forms that look different but are the same, based on their number of holes. From a visualization perspective, we may be disappointed that we did not get a donut or teacup; however, from a harmonic perspective, we got exactly what we asked for. *When a visualization is expressed as a feeling tone, the feeling tone can be expressed as a geometric pattern or as network unified by holes or nodes.*

Recognizing the harmonic data of a feeling tone search in real life can be challenging if we do not have good data evaluation methods. It is easy to visualize but is work to evaluate, organize, and act on data. The process requires a body/mind shift that can involve thoughts, judgments, emotions, beliefs, and body alignment. To put it another way, to have something new we need to let go of something old. **A vision sparks the Values Visualization Process, but the primary evaluation methods are values and feeling tone.** This is why it is so important in the beginning to consciously identify the values and feeling tone of a visualization.

CREATING INSTRUMENTAL AMBIENT SONIC FIELDS

You don't have to be a musician to create ambient sonic fields for yourself or others. A knowledge of music, music theory, and reading music is not necessary, and can even be an obstacle. Tapping tuning forks, sounding gongs, and playing crystal or metal bowls is easy to learn. However, all instruments require basic sound-making skills in terms of controlling volume, pitch, speed, and durations of the sounds being played. And, no matter which instrument is chosen, listening skills and a feel for sound and space are required.

Creating an ambient sonic field for yourself can be as simple as tapping a tuning fork, a crystal bowl, or a gong. In 2005, I developed the Awakening Bell in partnership with jazz great Jack DeJohnette and Gary Kvistad, the founder of Woodstock Chimes. Our goal was to create a universal sound-healing instrument that anyone could play instantly. The Awakening Bell is easy to play and is a very efficient way to work with the Values Visualization Process. *When the Awakening Bell is sounded, the long overtones of the bell create a theta-tuned ambient field of sound. A theta brainwave state is characterized by deep calm and increased creativity.* Using it is as simple as being aware of your intention when playing it and then tapping it. The whole process is effortless and can happen in a few seconds.

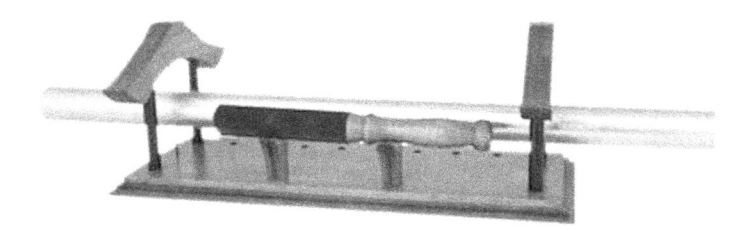

HOW TO CREATE AN AMBIENT SOUND HEALING CONCERT

A sound healing concert happens when you create an ambient sonic field for others. Today's sound healers use terms like Sound Bath, Sound Immersion, or Sound Healing Event to describe sound healing concerts. I call my tuning fork sound healing concerts "Human Tune Ins™" (see Appendix B). I met with the composer John Cage in 1978 in New York City. I had just had my work featured in *Ear Magazine*, a new music publication, which had just released its first Music Healing edition. John Cage was interested in the healing properties of music. He asked me if there was a difference between a music performance and a healing music performance. I said that all music is healing and that the answer is yes. There was a long moment of silence and he said, "Thank you."

I had no idea what I had just said to him. It just "came out" in the presence of a great composer whom I admired and respected. His teaching took place in a timeless silence between my answer and his thank you. It has taken me thirty-five years to begin to understand the answer I gave to John Cage. The answer is always evolving. The silence has never changed.

The performance of a sound healing concert is similar to and yet very different from a normal musical concert. The similarities begin with the basics. Regardless of the instrument being played, the performer needs to know how to play their instrument and control its dynamics within the concert space. They must be confident with their ability to play, listen, and become one with the sounds they are creating. This is called "getting inside the sound." Mastering their instrument, being inside the sound, and awareness of space are universal abilities required of anyone working with sound and music during a performance.

The difference between a sound healing performance and a musical performance is how the performer focuses on their audience and their formal music training. In formal music training, the emphasis is on playing well and pleasing an audience. Very little emphasis is given to the needs of an audience beyond the performer wanting them to appreciate what they are playing. Healing may happen during a musical performance, but it is not the primary focus of the performer. For example, in order to perform a classical piano concert, a pianist studies and practices for long periods of time to prepare. Once on stage, the pianist focuses on bringing a composer's music to life for their audience.

By comparison, the primary purpose of a sound healing concert is to empower the listeners so they can have a healing response. A sound healer specifically trains to enter into an intuitive empathetic relationship with their listeners. Everything about a sound healing concert—the concert space, how the audience listens, and what instruments are used—is to empower healing. A sound healer may or may not have formal musical training. What is most important is that the sound healer is masterful in creating sounds, getting inside these sounds, and being able to make intuitive, empathetic contact with his or her listeners. This requires a very different training than what is traditionally offered in a music school curriculum.

A sound healer's focus is to embed the sound waves with healing intention based on intuitive audience feedback. When a sound healer plays sounds within the network of their listeners' energy fields, the tonality of the sound naturally changes and becomes congruent with the energy dynamics of their listeners. The composer Dane Rudhyar called this "The Magic of Tone."[5] Being simultaneously inside the sound and bonded with their listeners, the sound healer will be naturally attracted to a specific listener or group of listeners. At this moment, they can tune into their needs at a universal archetypical level, and create a sound with embedded healing intentions just for them.

Listening to a sound healing concert is different from listening to a normal concert. During a sound healing concert, within the limits of the space, you can sit, lie down, and/or move around. Allowing your body to move is very important. This is different from a concert where you are expected to sit still and listen. In a sound healing concert, you should be open to allowing your body to move. Perhaps your body movements will be associated with unresolved experiences. In this case, it isn't necessary to remember the details of an experience to resolve it. You can just be in a dreaming trance state, allowing your body to move in resonance with the sound. Just focus on the sound in the light for the highest good and allow your body to unwind.

There are three types of movement that may take place during a sound healing concert: unwinding movements, communication movements, and sonic anchor movements. They are described below.

Unwinding Movements

Your body is intelligent; it holds somatic memories. Ideally, your body should naturally rise and fall with the sound waves. When something in you resists the waves, you may experience it as a pressure, as an uncomfortable feeling, or as an emotion or a thought. When this happens, let your body know that you are safe and then gradually let go of conscious control of your body. Letting go doesn't mean completely letting go. It means gradually letting go until you come to a place that is safe and your body is spontaneously moving. This is called unwinding.

To better understand unwinding movements, imagine that areas of pressure or constriction are part of your life that are "wound up like a rubber band." As you become aware of them and gradually relax, like a rubber band being freed up, you will naturally unwind. Unwinding movements should never be forced or planned. Just relax, trust your body and its intelligence, and allow it to naturally unwind. Normally, unwinding movements are unconscious micro-vibrational movements that you would not consciously notice. However, if you become aware of your body moving, you may notice a release, a slight pop, a twitch, or a deep breath. You may want to slightly shift your body, lift a shoulder, gently twist, rotate a leg, or you may want to stand up and make larger movements like walking, spinning, bending, or twisting. Be aware of the people around you and adjust your movements so as not to interfere with those around you.

Communication Movements

True spiritual voices are heard when the mind is in a state of deep absorption without conscious thought. While being in the sound, you may meet and have conversations with different beings. These beings are energy forms that appear in ways that are congruent with your reality structures. They can be parts of yourself, or of family members, people you know, or of people you don't know you are meeting in sonic space in the form of light beings, spirit guides, power animals, divine presences, or angels. When you communicate with others in normal conscious reality, your body and hands will naturally move to express something. The same sort of movement can happen during a sound healing concert. Although listeners may not speak out loud, it is OK for them to subvocalize. In other words, their lips and hands may move, but the sound is internal.

Sonic Anchor Movements

During a sound healing concert, you may want to bring your hand or hands to a specific place on your body. For example, if you want to work with your heart, place one hand or both on your heart. This is called a somatic anchor. Somatic anchors focus the sound waves moving through your body. After the concert, at any time, you can use your somatic anchor to recall the concert and to focus the sound vibrations on a specific area. (See Appendix C for how to create somatic anchors.)

TUNING YOUR BODY MOVEMENTS IN THE CONCERT SPACE

Be aware that every concert space has movement limitations. Be aware of and considerate of those around you when moving. It may be that you'll have to program yourself to make smaller movements or to bring your normal conscious mind back for a moment to find a safe place in the room to move. If it isn't possible to move around the room, then allow your body to make movements within your safe personal space that is defined by the room's limitations.

EMOTIONS

As the sound moves through your body, you may experience emotions. The term "emotion," from a sound healing perspective, means e-motion or energy-in-motion. Rather then focus on the emotion or on stories associated with the emotion, focus on the felt feeling of the emotion as a wave moving through your body. The sound healing space is not designed to support emotional outbursts or talking about your feelings. Therefore, it is important to focus on the wave of the emotion as a felt feeling in your body. It is up to the listener to set a boundary with their own emotions and not to allow its expression to interfere with other listeners. This is honoring your emotions in the right balance.

SUMMARY OF SOUND HEALING AND MUSIC CONCERT SIMILARITIES AND DIFFERENCES

Sound Healing Concert	**Music Concert**
Being present	Being present
Confidence with instrument	Confidence with instrument
Awareness of space	Awareness of space
Immersive listening	Listening
Musical training not required	Musical training required
Creating neutral ambient sounds	Performing a style of music
Listeners	Audience
Embed sound with healing intention	Performance
Intuitive empathic relationships	Playing music/entertainment
Sound healers are catalysts that disappear	Performers are center of attention

Five Element Sonic Fields

...and the beat goes on... — Sonny and Cher

Five Element Sonic Fields are created by sounds with a consistent beat. Underlying the field of changing life behaviors, there exists a unified and consistent life beat. Beats are invisible and yet without them we could not live. A heartbeat may range from slow to fast but, in order to live, it must be a steady beat. When a heart does not have a steady beat it is called arrhythmia, a condition which may be life threatening. In other words, we cannot fully live unless we are "on beat."

In the diagrams below, beat is represented by the black dots and changing life behaviors are represented by the dashes and slashes. Life behaviors change while the beat remains constant.

-- /	--- / __ ---///__	-	-- / __ -	__ ----//__	- /-
•	•	•	•	•	•

Even if the beat speeds up, it is still constant.

-- /	--- / __	---///__	-	-- / __ -	__ ----//__	-	/-
•	•	•	•	•	•	•	•

The events in our life and different movements of our body are the dashes and slashes. Even when our body movements and the events in our life appear to be random, they are unified by an underlying beat. When we successfully meet the mental, emotional, and physical challenges of our life events, we feel or sense the invisible beat that has been there all along. The rhythms in our life will always vary but the beat will always be consistent and unified.

Beat is how musicians, dancers, and athletes perform complex feats and make them look easy. When we try to imitate their moves, it's awkward because we are coming from the outside in. When we finally get it, we often say, "it was effortless," "it just happened," or "it feels like I didn't do anything." An accomplished musician will miss a note and never lose the beat. The innovative piano master Abby Whiteside used the term "emotional rhythm" to describe the ability to keep the underlying beat of music continually moving forward. She invented a number of creative exercises to tune musicians into the feeling of emotional rhythm. For example, she believed the eyes were externally focused and could easily disassociate us from the inner beat of the music. To remedy this, she would instruct pianists to play blindfolded in order to help them feel the continuous inner beat that transcended the rhythm of a composition.

When an accomplished athlete misses a beat, they say that they lost their timing. They know how to relax and tune back into the beat in order to rediscover their timing. In golf, for example, a beginning player will hit an errant shot and lose their timing. Instead of relaxing in order to find their inner beat, they will get tighter and continue to hit errant shots all day. In life, when we lose awareness of our inner beat and try to get it back, rather than learn how to relax to rediscover it, we use emotional words and phrases to describe our experience of being off beat.

"I feel scattered today."
"What a crazy day."
"I'm just not with it today."
"What a messed up day."
"Everything is just out of sync."
"Today is confusing."
"Pure chaos."

Like an athlete or musician, we naturally seek to get back on beat. When we are on beat, we have more energy and a greater ability to adapt. There is a law in systems theory which states the following: the least amount of diversity a system has, the more energy it conducts. Think about this in terms of rhythm and beat. Rhythm can have a large amount of diversity compared to beat. When we get lost in diversity, we lose our sense of beat, resulting in a dissipation of energy. When we are on beat, we experience the whole and we have unity in diversity.

The concept of returning to beat is the basis of many approaches to natural healing and spiritual renewal. When we are lost in the multitude of life rhythms, we naturally want to return to a simpler life and/or go back to nature. One way to do this is to eliminate complex life rhythms, making our inner beat easier to rediscover. For this reason, nature cure clinics are located in the countryside and eliminate complex life rhythms by implementing a simple and consistent schedule. People arrive scattered and out of sync and willingly participate in a protocol of prescribed activities. Following these activities is simple and "mindless," allowing the participant to relax and entrain to a consistent beat.

In a similar way, religions use ritual to bring people into a unified beat to renew their spirit. Rituals simplify and/or eliminate rhythm in order to emphasize the beat. From a vibrational perspective, a ritual is a set of behaviors performed in a cyclic manner which accentuates a precise beat. The activities within the ritual are kept simple and always happen on the beat. Whether the ritual ceremonies are Christian, Jewish, Buddhist, Muslim, or something else, they all have in common the establishment of a consistent and unifying beat.

Learning to perceive the elements in music and sound is a process which involves listening to the qualities of sound, i.e., beat, pitch, speed, volume, and rhythm, and becoming more aware of our physical sensations while listening to the music, as well as in trusting our intuition. Our experiences are then reviewed and translated into the energetic language of the five elements. The first and most important quality of the elemental carrier music is beat. *Webster's Dictionary* defines beat as a "pulse or throb" and defines pulse as "to vibrate or quiver." To understand the meaning of beat, we must merge with a piece of music and experience how the music moves us. Our body must be relaxed and we must be willing to feel the beat from the inside out. Beat is perceived kinesthetically; it is not an intellectual process. Our body must be free to move and then our mind can become aware of how it is moving.

In science, the concept of merging with the beat of the music is termed "mutual phase-locking of two oscillators" or "entrainment." It is a universal phenomenon. Whenever two or more oscillators in the same field are beating at nearly the same time, they tend to "lock in" and begin beating at exactly the same rate. The relationship of beat to the elements is given in the table below. To tune into the beat of your visualization and let your body move with it from the inside out, let your arms, hands, legs, and whole body come into a synchronized movement with the beat. Freely feel the beat until you perceive in yourself a whole-body response, a sense of felt resonance with the beat of the element.

In this sense, the four elements of Air, Fire, Water, and Earth represent different speeds of life beats that give rise to different categories of behaviors.

Element	Beat
Air	fast and light, erratic, quick
Fire	moderately fast, driving, staccato
Water	medium slow to moderate, flowing
Earth	slow to very slow, structured

The fifth element is Ether which can be visualized as a node or silence. Ambient sonic fields tend to be more etheric because they don't have a consistent beat. In order to intuitively understand the five elements, visualize a car with four gears and a clutch. Ether is neutral. It's like when we disengage the gears of a car with the clutch. Earth is first gear. Water is second gear. Fire is third gear. Air is fourth gear. In order to travel from one place to another, we must continually shift gears. Each gear is optimal for different speeds. Earth is for slow speeds. Water is for slow to moderate speeds. Fire is for a burst of fast speed. Air is for fast speeds. In order to shift in and out of different speeds, we need neutral. Without neutral, the gears would grind and we would be stuck in a gear.

Each gear is assigned to a different elemental behavior. When traveling in Earth gear, we will be slower and more organized, in control, safe, and secure. When traveling in Water gear, we flow between moderately slow to moderately fast speeds and enjoy the feeling of the ride. When traveling in Fire gear, we speed up to pass another car and are motivated by the burst of speed to reach our goal faster. When traveling in Air gear, we drive faster and think about all the things we are going to do when we reach our goal and at the same time listen to music and talk with friends on our Bluetooth. Ether gear is neutral. When we stop at a stoplight, we disengage the gears, go to neutral, and become still. When we push in the clutch while moving, we disengage the gears and enter neutral in order to shift between gears. Through neutral, we have access to all four gears at any time.

The Five Element Sound Dynamics Chart on the following page is a summary of behavioral and sound qualities that are associated with each element. Once a feeling of beat is achieved, the next step is to listen for these qualities of speed, volume, and pitch.

Sound healers don't have to be musicians; they can learn to cultivate the elemental qualities of music in the Values Visualization Process. If the music is associated with positive memories, it will be neutral and bypass the amygdala. Therefore, in choosing music, it is important to know a person's musical likes, dislikes, and associations with different styles of music. When we hear music associated with life experiences, the memories of those experiences immediately resurface—it can feel like they are happening all over again. Hippies who attended the Woodstock music festival replay their favorite festival songs to feel young and to relive their Woodstock experience. I grew up playing classical music and Broadway show tunes on the piano for my parents. Playing this music always brings back different childhood memories that are associated with different compositions.

FIVE ELEMENT SOUND DYNAMICS CHART

	Ether	Air	Fire	Water	Earth
Speed	• still	• very fast – fast	• moderately fast • moderate	• moderately slow	• slow • very slow
Volume	• silent	• erratic • intensity	• loud • moderately loud	• medium loud	• low – very low
Pitch	• silence	• very high • high	• moderately high • medium	• medium low	• low – very low
Quality	• space	• breathy • light, jumpy	• sharp staccato • driving	• flowing • connected	• repetitive • organized
Voice	• silent	• high and fast	• loud and staccato	• flowing • connected	• low • structured
Voice Content	• space around words	• skipping topics • multiple topics	• direct command • demanding • inspiring	• flowing without boundary • babbling on	• organized • matter of fact
Felt Beat	• still	• fast & light • expansion	• pounding • stomping	• side to side • sensual	• slow & even • structured stride
Intervals	• octave	• 4th & 5th	• 3rd & 6th	• 2nd & 7th	• tonic
Brain Wave	• gamma	• beta	• alpha	• theta	• delta

Music that we enjoy listening to bypasses the amygdala and enters the hippocampus of the brain. The hippocampus is part of the limbic system and consists of two nerve centers that curve back from the amygdalae. It records and stores memories and is responsible for our ability to remember events in time and space.

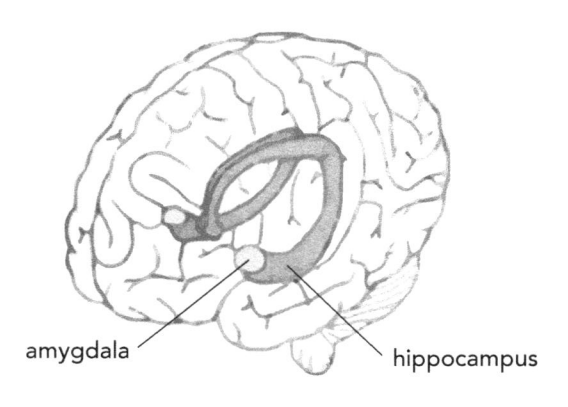

The hippocampus is important for sound healers because sound and music can be used to activate memories stored in the hippocampus which can have a positive healing effect. The late Dr. Oliver Sacks presented a video of a man in a nursing home suffering from dementia.[6] He is lethargic, depressed, and unable to communicate in a meaningful way. A therapist places headphones over his ears, allowing him to immerse himself in a blues music recording. The moment he hears the sounds he becomes animated, vocal, and able to verbally communicate. This happens because he has positive past memories of blues music; the sounds are able to bypass his amygdala and go directly into his hippocampus, where they wake up old memories and neural pathways associated with those memories. The result is an instant change in his personality and physiology. Dr. Sacks makes a connection between these instant psychological and physiological responses based on the ancient alchemic understanding of music as a "quickening art."

MUSIC AND SOUND ELEMENT INTERPRETATION PROCESS

The music and sound interpretation process is given in a numbered sequence for learning purposes. In practice, we should always go with what attracts us first when listening to the music.

1. Describe the first thing you notice. For example, is the music loud or soft? Is it fast or slow?
2. Tune into your body and feel the beat. This is normally very easy to do with Fire and Air music and more difficult with softer Water and Earth music. Generally, if you are having trouble tuning into the beat, the music is most likely Earth or Water or a combination of Earth and Water. To help get a better feeling for the elements, I recommend you experiment with the Element Hand Shaking Exercise in Appendix D.
3. Separately tune into each quality: speed, volume, and pitch.
4. Tune into the most dominant dynamics of the music. Use your felt body sense and intuition to translate the most dominant dynamics into an element or elements.
5. Tune into the music as a whole and translate your felt body feeling into an element or combination of elements. To refine your ability to move with the elements, I recommend *Moving with the Elements* by Thea K. Beaulieu.[7]

VISUALIZATIONS AND ELEMENTAL INTERPRETATIONS
Method 1

Tune into the feeling tone of what you visualize and match the beat of the sound or music to the feeling tone. We intuitively do this all the time without thinking. In the movies, in a love story, when a man wants to tell a woman he loves her, he could just say, "I love you." However, if he really wants her to get it at a deep level, he hires a band to play their favorite music. She relaxes into the sound and at that moment he says, "I love you." In order for the sound to be effective, he must know her, be clear about his intention, and then listen to different kinds of music in order to choose "the right sound." When his knowledge of her, his intention, and the music come into resonance, his body language becomes animated and he "just knows it's the right sound."

There are many situations where we choose sounds and music based on the elements. It's a big business. The advertising industry spends millions of dollars researching music and sounds that activate the element or elements they want to associate with their product.[8] Advertising market research includes product surveys, interviews, and systematic market testing in order to fine-tune the sounds and music used in their advertisements. They use sound-based terms like "your brand's voice" and "finding your brand's tone" to describe the integration of sound with their advertising messages. For example, an advertisement for Harley-Davidson motorcycles uses the deep sound of a Harley accelerating to the music of the rock band Steppenwolf's song "Born to Be Wild" (*Fire*). In contrast, an advertisement for Viagra uses the sounds of waterfalls and soft flowing romantic music (*Water*). **The advertising industry encodes element carrier sounds with the feeling tone of their products in order to make a sale. In contrast, sound healers encode element carrier sounds with the feeling tone that resonates with their client's or listener's highest values.** You can refer to Appendix E for web examples of the elements and advertising commercials, and Appendix F for Hypnosis and Element Integration Examples.

Method II: How to Create Sonic Fields Congruent with the Values Visualization Process

The beat of each element is like an archetype that gives rise to different categories of physical, emotional, and mental behaviors as seen in following chart.

FIVE ELEMENT BEHAVIORAL CHART

	Ether	Air	Fire	Water	Earth
Movement	• still	• fast, jumpy • erratic • defined	• strong • powerful • overbearing	• flowing • buoyant • heavy	• slow • frozen
Thought	• still	• wishing • "if only"	• get it done • "just do it"	• being with	• perfection
Quality	• deep calm	• wanting • lots of ideas	• pushing • power	• closeness • holding on	• critical of order
– Emotions	• grief	• detachment • remote	• anger • resentment	• sadness • jealousy	• fear
+ Emotions	• letting go • surrender	• compassion	• forgiveness • enthusiasm	• bonding • closeness	• courage
Story	• loss • letting go	• overload • lots of projects	• arguing • power struggle	• possessiveness • fantasy • loyalty	• completion • stuck
Expression	• giving space	• diversity	• motivation • will power	• creativity • sexuality	• order • cleaning up
Touch	• holding • space	• light • fast, quick • multi-directional	• direct • penetrating • focused	• kneading • flowing	• strong • secure
Body Structions	• throat • joints • ears • thyroid	• nervous system • skin • lungs • thymus	• heart • liver • adrenals • stomach	• generative organs • lymphatic system • secretory glands • bladder	• bones • colon • testes
Sense	• hearing	• touching	• seeing	• tasting	• smelling

This chart can be used in many different ways. I have spent over 45 years evaluating patients with element profiles. I am always discovering new ways of understanding how the elements mix together to create different behaviors.

The chart can also be combined with your intuition and initial behavior observations and used as a beginning guide to element sounds and music that can create sonic fields congruent with the Values Visualization Process. When the elements within a visualization are evaluated, the next step is to determine how to create the element balance. The element sounds or music should be in resonance with the element balance. The following case stories are designed to illustrate this in practice.

EDWARD

Edward came into his session excited about joining a health club. He was watching television and saw an advertisement about people working out, looking fit, and having a great time with other people working out. During the advertisement they played the Bee Gees song "Stayin' Alive" (*Fire*) and everyone was in rhythm and working out to the beat of the song.

Edward said, "I'm sitting around too much (*Earth*) and I'm thinking about joining a health club. I need to be more active (*Fire*). I know it would be good for me on so many levels."

I asked him, "How would your life be different if you were to join a health club and be more active?"

He replied, "I would feel better about myself. Getting out and moving is important to me. I know I would sleep better (*Earth*), and would be more healthy in general. Plus I would meet new people with similar interests."

I said, "So you really want to move more, exercise, and be more motivated?"

He said, "Yes, that's it."

I said, "And when you move more, you'll feel better, and more healthy?"

He said, "Yes."

I said, "And you'll meet new people and share your experiences?"

He said, "Yes."

Edward's values are being healthy, exercising, and being in a community that shares his values.

I said to Edward, "Close your eyes and imagine that you are at the health club. You are moving from machine to machine, stretching, and having a good time (*Fire*). Afterwards, you cool down (*Earth*) and have a healthy drink with someone you have met while exercising (*Water*)." I then asked him to tune in and tell me what he was feeling in his body.

Edward said, "I feel relaxed and alive. My body is warm and humming (*feeling tone*). (*Fire balanced with Earth and Water*)"

Fire in balance with Earth and Water is Edward's feeling tone. The visualization process uses his desire to join a health club to arrive at the feeling tone of what is really important for Edward. The more ways he has of achieving what he wants, the better. And the more he consciously understands how to value-test what he is looking for, the better his chances are of his continuing success.

Edward's story also illustrates the differences between advertising intention and the Values Visualization Process. Advertisers focus on how to get Edward to buy their products. The Sound and Visualization Process uses Edward's desire to join a health club to discover many different ways of expressing his values. An average of 80% of people who join health clubs to express their values of motivation, exercise, health, and community quit going after five months, and only 18% of the remaining members use the health club on a regular basis.[6] The odds are not in Edward's favor if he joins a health club as the only expression of his values.

MARY

Mary said during her session that she wanted to find her life partner. I asked her to visualize and describe being with her ideal life partner. She said, "I see a handsome, tall, and caring man (*Water*) who is active (*Fire*) and likes to do many different things (*Air*). We are able to be together and share our life experiences and openly discuss our life challenges (*Water*). We are able to give each other space (*Ether*) and come back together and share (*Water*). We are always doing fun things and going places together (*Air and Fire*). Most importantly, we are faithful and always look forward to being together, expressing our sexuality in many different ways, and we are always sharing (*Water*). He is someone I can trust and we always acknowledge each other (*Water*)." Mary was quiet for a moment and then she added, "And, oh yes, he is financially stable. He doesn't have to be rich. He just has to have a realistic sense of finances. (*Earth*)"

TOM

Tom said during a session, "I have lots of ideas (*Air*) and my goal is to write a book."

I asked, "Can you visualize yourself writing?"

Tom said, "I am sitting in a quiet space (*Ether*) surrounded by nature (*Earth*). I am trusting in my ideas (*Earth*) and writing them as they come (*Air*). My ideas are flowing and I am into it (*Water*). Sometimes I just get an idea and it bursts out and I am excited (*Fire*). When I have pages of ideas, I see myself connecting them (*Air*) and organizing them (*Earth*) into a book that's very important to me on a deep level (*Water*)."

MATCHING MUSIC AND THE ELEMENTS

Ether, Air, Fire, Water, and Earth can be visualized as sonic energies that combine in many different ways. A piece of music is rarely a single element, although a single element may predominate. **The goal is to, as best as possible, choose a music**

that comes closest to the elemental qualities of a visualization. The following are examples of different elements in different styles of music.

ETHER
  John Cage: 4'33"
  John Beaulieu: *Calendula (Harmonic Clouds)*
  Philippe Pascal Garnier: *The Healing Sounds of Crystal*

AIR
  Ernesto Cavour: *Rio Cansado (solo de Charango)*

AIR WITH FIRE
  Ernesto Cavour: *Matraca de Quena y Charago, Rosario de Uvas*
  Poulenc: *Concerto for 2 Pianos D minor – Allegro ma non troppo*
  Vangelis: *The Motion of the Stars*

AIR WITH WATER
  Ravel: *Daphnis et Chloe, Suite No. 2*

AIR WITH EARTH (FAST AND PRECISE)
  Mozart: *Sonata No 10 C Major Allegretto*
  Infected Mushroom: *Never Mind*

FIRE
  Steppenwolf: *Born to Be Wild*

FIRE WITH AIR
  Ludwig Beethoven: *Piano Sonata #23 in F minor, Op 57, "Appassionata" – 3, Allegro Ma Non Troppo*
  Khachaturian: *Toccata*
  Bach: *Brandenburg Concerto No. 1 in F Major*
  Junior Brown: *Highway Patrol*

FIRE WITH WATER
  Beach Boys: *Surfin' USA*
  Metropolis: *Ecaroh Turtle Island String Quartet*

FIRE WITH EARTH
  Steve Reich: *Sextet: 3rd Movement*
  Eurythmics: *Sweet Dreams Are Made of This*
  The Pretenders: *Don't Get Me Wrong*

WATER
  Enya: *Watermark*
  Beaulieu: *Moonzak*

WATER WITH EARTH
  Henry Mancini and His Orchestra & Chorus: *Moon River*
  Beach Boys: *Surfer Girl*
  Beethoven: *Moonlight Sonata Op 27 No. 2*
  Louis Armstrong & Ella Fitzgerald: *Dream a Little Dream*
  Patsy Cline: *I've Loved and Lost Again*

WATER WITH FIRE
  Incendio: *Maranga*
  Sade: *Smooth Operator*
  Kathy Mattea: *Eighteen Wheels and a Dozen Roses*

WATER WITH AIR
  Frédéric Chopin: *Nocturne #12 in G Op 37/2*
  The Chordettes: *Mr. Sandman*

EARTH
  Tibetan Tantric Choir: *Guhyasamaja Tantra, Chapter II*
  Phillip Glass: *Koyaanisqatsi*

EARTH WITH WATER
  Erik Satie: *Gymnopedie No 2*
  Stravinsky: *Rite of Spring, Part I: Adoration of the Earth*
  Silvia Nakkach: *Shaman Journey*
  Dario Domingues: *Barco De Papel*

EARTH WITH AIR
  Johann Sebastian Bach: *Sonata in E Flat Major*
  Claude Debussy: *Syrinx pour flute seule, L.129*
  Deuter: *Sky Beyond Clouds*

EARTH WITH FIRE
  Taiko Drum Ensemble: *Miyake Daiko*
  Phillip Glass: *Mishima*
  Bob Marley: *Red Red Wine*
  Clint Black: *Put Yourself in My Shoes*

SECTION THREE

Sound Healing and Values Visualization Practice

How to Use Sound to Work with Dependencies

For purposes of this chapter, a dependency is defined as a manageable behavior that contributes to our life and that has gotten slightly out of control. For example, drinking a cup of coffee in the morning can be helpful; however, drinking coffee may become a dependency when you start to drink more cups throughout the day and have trouble sleeping at night. Another example of a dependency is chocolate. When you eat a piece of chocolate every now and then, it's usually an enjoyable behavior. Eating chocolate can become a dependency when you eat more chocolate in order to deal with stress.

An addiction is a behavior that results in your life becoming unmanageable and out of control, to the point where it can become hurtful to yourself and others. Working with addictions requires support that may include counseling, medical care, and a community such as a twelve-step program. If you are working with a counselor, a doctor, and/or are in a twelve-step program, this sound healing work will integrate with and enhance your healing process. If you have an addiction that is out of control and you do not have a counselor, I suggest you seek help and then use the following sound healing protocol for dependencies as part of your recovery process.

Working with a dependency using sound healing begins with understanding the value gained from the dependency. This may seem counterintuitive. Our reaction to dependencies and addictions is to assume they have little or no value. For example, if someone is smoking, our first impulse is to tell them how bad smoking is for their health. Anyone who smokes can see this written on every pack of cigarettes they purchase. One would think that with all the research about the bad effects of smoking more people would stop smoking.

The reason people continue to smoke is that they get value from smoking. If they were to stop smoking, they believe, either consciously or unconsciously, that they will lose access to the value they receive from smoking. Therefore, in order to stop smoking, one must become conscious of the values gained from smoking. One must be open to understanding and creating new ways to obtain the same value and/or better value than by smoking. When you have behaviors that are as good as or better than the value achieved by smoking, you will naturally move into a more efficient and healthier way of living those values.

In 1971, it was discovered that 40% of servicemen in Vietnam had tried heroin and that nearly 20% of them were addicted. Researchers thought that when they returned home there would be a real problem with addiction in America. In response, the Special Action Office of Drug Abuse Prevention (SAODAP) was created to track veterans.[1] They found that when the addicted soldiers returned to America, 95% of them eliminated their addiction nearly overnight. The findings by the SAODAP completely contradicted the patterns of normal addictions. The normal addiction cycle would require that an addicted user enter a clinic and get clean. After normal treatment, the readdiction rate is 90% or higher. Nearly every heroin addict relapses, whereas 95% of those returning from Vietnam just stopped using heroin and got on with their lives with minimal to no withdrawal symptoms or relapses.

One of the explanations for the findings given by psychologists was that there was great importance given to changing their environment. Simply removing them from an environment that triggered all their old habits made it easier to break bad habits and build new ones. Changing environments is important. Of equal importance, however, is that the value of the new environment supports the positive outcome of the dependency. The sound healing dependency protocol focuses on understanding and using the value gained from a dependency as a tool for motivating values focused on environmental change. Ideally, the value and sound healing work would be part of a process that supports new positive value expressions within a new environment.

A mistake we often make is associating value with the dependency delivery modality. For example, chocolate is made from cacao, the ceremonial drug of the Mayans. When used in the right setting, cacao is a proven and effective delivery mechanism for entering into an altered state of consciousness. Once we learn the experience and the value this state brings to our consciousness, we can begin the process of discovering new ways to express it. Indigenous shamans believe that cacao is a plant teacher, and that we need to honor our teacher in order to own what we have learned. Only through honoring the teacher can we say goodbye to the teacher. When this happens, we are no longer dependent and we enter the process of discovering new ways that are congruent with our life and wellness to manifest the values we have learned.

The understanding by shamans that substances such as chocolate, marijuana, alcohol, tobacco, opium, mescaline, ayahuasca, opium, etc., are teachers is helpful in explaining why different dependencies are easier to let go of than others. We all have had different teachers in our lives. Ideally, teachers are like catalysts. They inspire us and then let us go. Some teachers, though, are harder to let go of than others, and vice versa—some teachers hold on to us longer than others. Each teacher has a unique personality that we may or may not be attracted to. The same is true with plant teachers and ultimately all dependencies, plant-based or otherwise. Chocolate, marijuana, alcohol, tobacco, mescaline, ayahuasca, and opium are plant teachers that have different personalities and different lessons to teach. Once we have learned the lessons, we no longer need to stay attached to the teacher.

SUSPENDING CRITICISM AND JUDGMENT

The first step in working with a dependency is to suspend your judgment and criticism of yourself for doing your dependent behavior. For example:

- I eat too much chocolate and this is bad for my health.
- Playing computer games all day isn't good. I must be a lazy person.
- I just keep drinking coffee. I know it's bad for me. I'm jittery all day and I can't sleep at night. I just have no self-control.

To suspend your critical and judgmental thoughts, first recognize that they have not helped you to stop your dependent behavior. Second, thank them for caring and ask them to step aside for the process to continue. Let them know that the process you are entering into is not to get rid of your judgmental voice. This may sound counterintuitive to the critical/judgmental self, but anything less is denying your creative intelligence.

CREATING SUCCESS

As strange as it may sound to the critical/judgmental voice, success is success. And success can be learned from and expanded into other areas of your life. It is better to start with a base of success than a base of failure and depression.

When a person stands before their twelve-step group and says, "I am an alcoholic," from this perspective they are saying that they are really good at drinking. Being an alcoholic does not make them a bad person. The things they did while drunk may be out of control and bad and part of the process is making amends for these things. It may be in their best interests to find behaviors other than drinking alcohol to express their success.

> A teenage client of mine spent hours creating a fake driver's license in order to purchase beer. He learned how to use Prestype, create exact lines and spaces, and, in general, how to focus on a demanding project. Today, he uses the same behavior to paint pictures and design books. He stopped drinking beer when he was 20, but if he ever needed to drink beer again, I know he'd be good at it.

One of the ways a dependency is different from an addiction is that an addiction requires amends for out-of-control and destructive behaviors. To accomplish this, the judgmental voice is needed in order to admit to themselves that they have used being good at their dependency to hurt others. Then they must take responsibility for their actions. From this perspective, alcohol is not the problem. The real problem is how to properly manage being good at their dependency. In other words, success is the real biochemical delivery system. The question then becomes which delivery mechanisms are optimal for the best outcome, or what is the best way to get what they want.

IDENTIFYING YOUR VALUES

These exercises help you discover the value you are getting from your dependency. Begin by assuming that your dependent behavior is an intelligent choice. To tune into its value, close your eyes, and then recall the experience of your dependent behavior. Some recollection questions that may help are:

- Is there a special time I do this behavior?
- Is there a special place I like to do this behavior?
- Is there a special brand I like for my behavior?

- How do I prepare for my behavior?
- What do I notice in my body when I do this behavior?
- What happens in my mind when I do this behavior?

Dependency Session Example: How to Identify Your Value

A woman taking my class said that her dependent behavior was overeating chocolate. She told us that she knew too much chocolate was bad for her and that she needed to do something about it.

I said that I knew that too much chocolate wasn't good for her but that I wanted her to focus on the value she got from eating chocolate and would this be OK? She agreed that it would be.

I explained to her that during the session nothing would be said or done to take away chocolate. I clarified that the goal of the session was to review the positive values achieved by eating chocolate and to discover new ways of achieving those values. I asked her, "Are you open to exploring new positive outcomes that are as good as or better than eating chocolate and finding ways to create them in your life?" She said she was.

I asked her to close her eyes and to see and feel herself eating chocolate. She said, "I'm relaxed. I feel like I'm alone and in my own space. In this space I get a sense of peace. I feel warm inside and I feel good about myself."

Next, I helped her explore the values in her description which she said were:
- Relaxation
- Being in and having my own space
- Feeling safe and at peace
- Feeling warm comfort

When a person is asked to recall the positive outcome of a dependency, they are accessing the outcome without actually doing the dependent action. When I did a recall process with this client in front of a class, everyone immediately experienced the positive outcome of the dependent behavior being worked with. The students were able to observe changes in the client's body, which included balancing of the shoulders and spine, a smile, a flushing of blood to the face, a deep relaxation, and pupil dilation. More importantly, they sensed a feeling tone that drew them into sympathetic resonance with the higher state of consciousness. Without eating chocolate, everyone arrived at a higher state of consciousness by one person's experience of recalling the positive outcome of eating chocolate.

Once the higher state of consciousness is identified by value and feeling tone, the next step is to let go of the teacher and claim the state for your own. I said to my client, "Chocolate has been your teacher. It has guided you into wonderful states. Are you willing to go into the sound and discover new ways that are as good as or better then the values the chocolate has taught you?"

She closed her eyes and said, "Yes."

I began tapping the tuning forks.

In summary, recall your dependent behavior, get inside of it, become it, and then step back and discover the value you get from your dependent behavior. Write those values down.

A QUESTION TO ASK YOURSELF OR THE PERSON/GROUP YOU ARE WORKING WITH

Are you open to different ways to express your values?

FEELING TONE PREPARATION

Close your eyes and tune into your body and the felt feeling of the positive values of your dependency. Let your rational mind go and tune into a deeper sense of vibration in your body. Let your body subtly vibrate like a sounding board, projecting your feeling tone like a wave moving through space.

SOUND HEALING WITH FIBONACCI TUNING FORKS*

For Dependencies in General

Tap C & G together and continue tapping and move with the sound. Let go of any rational thoughts and "get inside the sounds."

For Dependencies in General

Hold C256 in one hand and G, A, and C512 in your other hand. Tap with the C and continue tapping and move with the sound. Let go of any rational thoughts and "get inside the sounds."

---

* C & G are the gateway into the Fibonacci spiral and can be effectively used with each dependency. The suggested additional Fibonacci tuning forks are explained in depth in *Human Tuning: Sound Healing with Tuning Forks.*

Chocolate, Ice Cream, and Sweets

Hold C256 in one hand and G, A, and 5/8 in your other hand. Tap with the C256. Continue tapping and move with the sound. Let go of any rational thoughts and "get inside the sounds."

Alcohol

Hold C256 in one hand and G, 5/8, 8/13, in your other hand. Tap with the C256. Continue tapping and move with the sound. Let go of any rational thoughts and "get inside the sounds."

Marijuana

Hold C256 in one hand and G, A, 8/13, in your other hand. Tap with the C256. Continue tapping and move with the sound. Let go of any rational thoughts and "get inside the sounds."

Hard Drugs, i.e., Opiates, Cocaine, and Amphetamines

Hold C256 in one hand and G, A, 13/21, and 21/34 in your other hand. Tap with the C256. Continue tapping and move with the sound. Let go of any rational thoughts and "get inside the sounds."

DEPENDENCY TRANCE

As you sit relaxed and comfortable, allow yourself to find the depth of relaxation that is just right for you to work with. Let all of your creative resources come forth now in order to create new pathways for expressing the positive outcome of your dependency.

Imagine that for a violin string to play the correct tone, that string must have just the right amount of tension. It can't be too tight or too loose. It must be the perfect tension to play the correct tone.

You have done a great deal of work clarifying the positive outcome of your dependency and the values you receive from that outcome. Allow the felt sense of those values to guide you into their resonant feeling tone. Let all thoughts go and become the feeling tone. Trust in it and let it be automatic and continuous.

As you listen to the sounds, allow your creative self to discover and create new neural pathways than the ones you have used in the past, that are as good as or better for expressing the values you have learned from your

dependent behavior in new and creative ways. Let yourself dissolve and create ways to express these values.

Tap Fibonacci tuning forks.

Working with Cancer and Health Challenges

The cancer protocol can easily be modified to work with heart disease, autoimmune disorders, digestive disorders, sexual disorders, and depression. When you see "cancer" just substitute the health challenge you are working with. The Values Visualization Sound Healing Process is based on the assumption that anyone with cancer (substitute any heath challenge for cancer) wants to live a life of value. Although we think of fighting cancer as receiving chemotherapy, surgery, or radiation, there is a much larger and often overlooked dimension. If you are fighting cancer, you might ask yourself, "When the cancer goes into remission who will I be? What will I do?" The answers to the above questions require a future visualization, but the feeling tone created by a future visualization is in the present. Research studies show that happy, joyful, loving people live up to ten years longer than unhappy people.[2] The thought of a future filled with love, joy, optimism, and happiness stimulates our body's ability to create oxytocin, dopamine, nitric oxide, and endorphins, all of which signal our body's natural cancer-fighting abilities.

VALUES VISUALIZATION PROCESS EXAMPLE

Visualize and describe yourself living in the future without cancer. Ask yourself these questions, "When the cancer is gone, what do I really want to do?" "How will my life be different?"

Next, look over your answers and determine your values. To do this, you will need pen and paper. Sit quietly or with another person or group that you trust. Begin by reviewing and/or sharing your vision and writing down the personal values contained in your vision. For example:

> **EXAMPLE 1:** When the cancer is gone, I'm going to walk out of here and play a lot of golf. I'm going to travel to different countries and play lots of new golf courses. I'm going to compete and have fun and afterwards sit with friends and acquaintances and talk, laugh, and enjoy good company.
>
> **VALUES:** Competition, achievement, coaching, adventure, friendship, fun

EXAMPLE 2: When the cancer is gone, I'm going to slow down my life pace and appreciate my family. I'm going to make time for family dinners and for going on family vacations. I'm going to go off on my own and have adventures and share them with my family.

VALUES: Family, patience, listening, caring, adventure, future generations

Once you have determined your values, always ask yourself, "Am I open to different ways to express my values?

Next, to tune into the feeling tone of your visualization, describe it in sensory words. For example:

I am at my studio and sunlight is streaming through the windows. I look at the paint on my hands and clothes and I feel like a rainbow. My paintings are everywhere and are illuminated by the sunlight. I can feel the textures of the colors with each breath. I hear the voices of my friends at the door. I open the door and everyone is happy and excited as they see my studio and paintings. Our conversation about painting is like a flowing river of sound. We're hungry and we eat sitting near the window. The taste of the food and the sound of my friends' voices merge into a feeling of community.

Once you have a sensory visualization, read it out loud and/or have someone you trust read it. During or after the reading, create sounds. For example:

- When the reading of your sensory visualization ends, knee or activator tap the C & G tuning forks, bring them to your ears, and be with the sound.

- Stand in a safe space. When the reading of your sensory visualization ends, take a deep breath and tune into the feeling tone of your visualization. Tap the C & G tuning forks together and allow your body to move with the sounds. Continue tapping and moving.

- First choose your target. For example, if you have lung cancer then let your lungs be your target. You can be as general or as specific as you want to be. For example, if you have had an MRI or have a specific anatomical location you want to focus your attention on, you can guide the sound and your feeling tone to that area. For example, sit in a safe space. Take a deep breath and tune into the feeling tone of your visualization. Knee or activator tap the C & G tuning forks and bring them to your ears. As you listen to the sound, visualize it going into and vibrating your target area.

TUNED HUMMING SESSION EXAMPLE

PREPARATION: Sit in a safe space. Take a deep breath and knee or activator tap the C & G tuning forks. Close your eyes and mindfully listen to the sound and hum with a sound that resonates. Remember your humming sound, take a deep breath, and then hum without the tuning forks. Next, hum with the tuning forks and, again, hum without the tuning forks.

SESSION: Tune into the feeling tone of your visualization. Imagine being inside the sound of the tuning forks and hum. Allow the humming to move throughout your whole body and, at the same time, you may choose to send the sound to a specific area. To create a circuit, place your hand on the target area.

If you are lying down, place an open hand, if possible, on the back of your neck where your skull meets the spine (the soft tissue just inferior to the occiput and over C2). Place your other hand on the target area, take a deep breath, and then hum. The hand at the base of your occiput helps to relax your vagus and trigeminal nerves; this will enhance the healing power of the sound.

Creating Your Perfect Living Space

This protocol can easily be modified for "a perfect room," "a perfect car," "a perfect vacation resort," etc.

STEP 1: Visualize and describe your ideal living space (room, car, etc.) in sensory-based terms.

STEP 2: Determine the values expressed by your new living space. Sit quietly, either by yourself or with a person or people you trust. Begin by focusing on and writing down the values you are expressing in your new space. For example:

CREATIVITY: I can create areas where my creative projects can take form and manifest.

SPACE: I am able to move around freely.

WARMTH: I feel a sense of myself. Everything that is in the space reflects a part of my inner self in a special way.

SHARING: I can invite friends to my space and have a good time sharing

STEP 3: Are you open to different ways to express your values?

STEP 4: Tune into the feeling tone of your visualization using sensory words. Imagine being in your perfect living space. Tune in to your body and to the felt feeling of being in your perfect space. If you are aware of your emotions, allow yourself to be with your emotion or emotions at a safe level. They sometimes come because you are on the right path. If you are having thoughts that are judgmental or critical, thank your thoughts and let them know you are going to be OK. Your thoughts and emotions are most likely trying to protect you from being hurt. You are not getting rid of or trying to disassociate from your thoughts and emotions. You just want to acknowledge them and put them into a different perspective.

You are now ready to get in touch with the feeling tone of being in your perfect living space. Allow yourself to tune into a deeper sense vibration in your body. Let your body subtly vibrate like a sounding board, projecting your feeling tones like a wave moving into space.

STEP 5: Visualization for Finding Your Perfect Living Space

SHORT VERSION

Before you go to sleep, tune into the feeling tone of your ideal living space.

Say either out loud or to yourself: "I ask my creative self to travel in my dreams tonight and look over all the dream living spaces that resonate with my perfect living space feeling tone."

Knee tap C & G tuning forks and/or tap them together, move around with the sound, and pretend that you are dancing in your new living space.

When you wake up, become aware of any physical sensations or felt feelings in your body. Meditate on that feeling, ask for a great day, and tap or knee tap your C & G tuning forks and bring them to your ears.

MEDIUM VERSION

Sit comfortably and find the depth of relaxation that is just right for you to work with. Let all your creative resources come forth now in order to find your perfect living home.

You can imagine that for a violin string to play the correct tone, the string has to have just the right amount of tension. It cannot be too tight or too loose. It must be the perfect tension to play a great concert.

Find that perfect level of relaxation, with just the right amount of tension, for you to do the work necessary to find your perfect living space.

You have done a great deal of work clarifying the values you want to experience in your perfect living space. Allow those values you learned about to guide you to the feeling tone of living in your right living space. Later you will need all of those values to evaluate potential right living spaces.

Now, become aware of your feeling tone, of being in your ideal living space.

Take your time and allow the feeling tone to vibrate throughout your body. If you are not consciously in touch with your feeling tone, then trust and ask that it vibrate your body at a safe level.

Let go of your thoughts and allow your feeling tone to continue in the background like a hum.

Tap C & G tuning forks or play a sound healing instrument.

LONG VERSION

Imagine you can have any living space you want that is congruent with the optimal expression of your life values.

You are in a huge dream store that is filled with millions of living spaces. Each one is a different size, shape, and look. As you enter the store, there is a sign that reads "Welcome to the Dream Living Space Store." Feel free to walk around, fly around, walk through walls, and transport effortlessly from space to space. You are welcome to manifest furniture and live in any dream space. In the dream store, time is measured in dream time. One second in real time can be days in dream time.

When you enter a dream living space and you want to get a better sense of living in that space, just adjust your dream time clock for that space. If you like one living space and want to live there for a little while, then adjust your dream time clock to give you a special signal to bring you back when you want to return to your normal dreaming time.

The dream living space store is enormous. In order to better explore the store, tune in to the feeling tone of your right living space. Ask to put the space in the light for the highest good and that your dream body be transported to that part of the dream store that has spaces that are in resonance with your feeling tone.

As your dream body materializes, feel free to look around. When you see a dream living space that "clicks" or "calls to you," immediately go inside.

When you enter the door, feel yourself in the living space. Ask your creative self to see yourself in this space successfully expressing your life's values.

Ask yourself: Is this space too big or too small?

When you get an answer, ask your dream store to change the walls to tune your space to the right size for you.

You are fine tuning the feeling tone of your perfect living space. You can get lost in a space that is too big and constricted in a space that is too small. In your dream store you can experiment with different space sizes until a size "clicks."

Go from room to room, tune the space of each room, and adjust it to fit you.

Tap C & G tuning forks or play sound healing instrument of your choice.

Come back to the entry room and tune into the feeling tone of the whole space. Let that feeling tone vibrate your dream body.

As the feeling tone vibration moves through your dream body, allow your dreaming self to dissolve into the field of infinite potential. Within the field, let the feeling tone attract and pull together fine creative energy strings. Let the feeling tone naturally weave them into a geometric seed pattern of your ideal living space. Do this now.

Tap C & G tuning forks or play sound healing instrument of your choice.

Leave the field of infinite potential and bring that special dream space seed pattern back to your dreaming self. Let it sprout and grow into a dream living space that is just right for you. Feel free to live in that space in your dreams every night. Feel free to safely daydream about your ideal living space. Listen and be with your ideal living space now.

Tap C & G tuning forks or play sound healing instrument of your choice.

As you come back to your waking self, feel free to create a fine dream neuron thread between your dream living space and your waking life. Keep that thread in a special place inside of your neural net. When you see or hear of a living space in waking life, the thread will vibrate like your phone and let you know where to look and where to find it.

Finding Your Ideal Life Partner

This protocol can easily modified for a ideal friend, business partner, teacher, etc.

STEP 1: Visualize and fantasize being with your ideal life partner and describe it in sensory terms.

STEP 2: Determine the values expressed by being with your ideal life partner. You will need a pen and paper. Sit quietly with a person or persons you trust. Begin by focusing on and writing down the relationship values that are important to you. For example:

SHARING: We are able to share life experiences and openly discuss our life challenges.

SPACE: We are able to give each other space to discover and express ourselves without the other needing to be present for everything.

TRUST: We can each go our own direction and trust that we are faithful and look forward to coming back together and sharing.

ACKNOWLEDGEMENT: We always acknowledge each other and treat each other with respect.

FUN: We do things together and have fun.

STEP 3: Are you open to different ways to express your values?

STEP 4: Tune into the feeling tone of your visualization using sensory description terms.

Imagine being with a partner who shares your values. Tune in to your body and to the felt feeling of being with a partner who shares your values. If you are aware of emotions, allow yourself to be with your emotion or emotions at a safe level. They sometimes come because you are on the right path. If you are having thoughts that are judgmental or critical, thank your thoughts and let them know you are going to be OK. Your thoughts and emotions are most likely trying to protect you from being hurt. You are not getting rid of or trying to disassociate from your thoughts and emotions. You just want to acknowledge them and put them into a different perspective.

You are now ready to open yourself to get in touch with the feeling tone of being with your ideal partner. Allow yourself to tune in to a deeper sense vibration in your body. Let your body subtly vibrate like a sounding board, projecting your feeling tone like a wave moving into space.

STEP 5: Visualization for Finding Your Ideal Life Partner

VERSION 1: DREAMING

Before you go to sleep at night, focus on the feeling tone of being with your ideal partner. Ask that in your dreams your dream body will search for resonant feeling tones in the larger field of shared dreams. **Imagine your feeling tone as a sonic frequency that can attract new relationships into your life that are resonant with your ideal feeling tone.** Be open to those relationships showing up in your life in different ways and forms. Finally, let that part of yourself that may have been hurt in the past know that you will value test every relationship that appears with your feeling tone resonance.

Let go of your thoughts and allow your feeling tone to continue in the background like a hum. Let it be automatic. Tap your C & G tuning forks on your knees and bring them to your ears. Trust and surrender yourself to the sound. You can switch the tuning forks from your right to your left hand, tap again, and then go to sleep.

VERSION 2

This can be done with a partner who reads it to you or you can create a recording that you can listen to.

As you sit comfortably, allow yourself to find a depth of relaxation that is just right for you to work with. Let all of your creative resources come forth now in order to find your life partner.

Imagine that for a violin string to play the correct tone, the string has to have just the right amount of tension. It cannot be too tight or too loose. It must be the perfect tension to play a great concert.

Find that perfect level of relaxation, with just the right amount of tension, for you to do the work necessary to find your life partner.

You have done a great deal of work clarifying your relationship values and allowing those values to guide you to the feeling tone of being with your ideal life partner. Later, you will need all of those values to evaluate potential life partners.

Now, become aware of the feeling tone of being with your ideal partner.

Take your time to allow the feeling tone to vibrate within your body. If you are not consciously in touch with your feeling tone, then trust and ask that it vibrate your body at a safe level.

Let go of your thoughts and allow your feeling tone to continue in the background like a hum. Let it be automatic and continuous.

Tap C & G tuning forks or play sound healing instrument of your choice.

Imagine that you have within you a vast ocean of knowledge and experience.

Further imagine, as in a dream, that you are capable of locating your creative self anywhere within that vast ocean.

Discover a special place where your dreaming self can meet other dreamers. And, even beyond that place, seek a special vortex that leads to a field of potential from which all dreamers come.

Allow yourself to move and to travel in your dream body through the field of dreamers to and through the vortex to the field of infinite potential.

Let your dream body dissolve into the field, and let the field begin dreaming you.

Without having to understand, allow the feeling tone of your ideal partner to vibrate within the field of infinite possibilities, like a tuning fork. Imagine that your feeling tone is signaling the creative substance of dreams to create a dreaming form that resonates with your feeling tone.

Imagine that your creative dream substance is signaling through multi-vortexes to other dreamers that resonate with your feeling tone.

Be with this for a while as you surrender to your dream body and let your dream body surrender to the greater field of infinite potential. Let your feeling tone vibrate and trust as you listen to the sounds.

Tap C & G tuning forks or play sound healing instrument of your choice.

Now it is time to leave the field of infinite potential and to re-enter your dream body with the feeling tone of your ideal partner continuing to hum in the background.

Within the field of dreams, allow your dream body to meet other dream bodies that are in resonance with your ideal feeling tone.

Let time expand. In your dream body you are capable of meeting thousands of dreamers in a few seconds or less. These microseconds can be like days.

You are a creative, dreaming being with a focus. You know what you are looking for. You are open to it manifesting in dream time and finding its way to real time.

You do not need to know exactly how your car runs to drive it. You do not need to know exactly how your computer works to use it. You can go online and type in a search word and it will search. **Imagine that your feeling tone is a search word used to search in the vast field of dreamers. Google your dream partner in your dreams.**

In your dream body, allow your creative self to look over all the possible matches.

Be open to those matches that are resonant with your feeling tone. Allow them to appear in your waking life in many different ways.

Tap C & G tuning forks or play a sound healing instrument of your choice.

Now, allow your dreaming self to move into the background and your waking self to gradually surface. As you awaken, remember to reaffirm your values.

Let your waking self know that it is important to use your values to evaluate potential partners that will appear.

Allow your creative self the freedom to look over the relationships of your past and to learn from them. Let your creative self do value checks and discover those values that attracted you to a partner and also the values that your partner had that were not congruent with your values.

Learn from your past relationships. Let yourself become stronger and more focused. Commit to your values.

When that learning is complete, allow your waking self to fully return to waking consciousness.

In your dreams tonight, continue to discover potential dream partners. When you wake up in the morning, feel those dreams in your body and allow your dream journey discoveries to integrate with your waking self, knowing that the signal has gone out to find your new life partner.

Winning: Visualizations for Success

This protocol can easily modified for all sports, music performances, taking an exam, etc.

Go to the store and purchase a lottery ticket. One lottery ticket should cost no more than two dollars. When you get your ticket, hold it in your hand and visualize what you would do if you won. Have fun doing this. Become aware that visualization is natural, much like breathing, and trying not to visualize will not work. I once gave a lottery ticket to a mathematician friend who gave me the odds of winning a million or more dollars. He said I had better odds of getting struck by lighting twice then winning the lottery. When I gave him the ticket he laughed. I told him it was his ticket, and he could throw it away. He paused, his eyes opened wide, and he smiled and said, "No way. I might win."

1. Hold your lottery ticket and share with someone you trust your fantasy of what you would do if you won the lottery. This works even better if the person you are sharing with has a lottery ticket as well and you both brainstorm what you would do if you won.

    For example, name two or more things you would buy if you won the lottery, for example, a new sports car, a yacht, a house on the beach, etc. You can use the exercise to work with everything you want. However, for this exercise, choose one thing you would purchase and then say what you would do with it. For example:
    - What would you do with your new yacht?

      I would cruise the Caribbean with my family and friends.

2. Tune into your future lottery winning purchase on a sensory level. List two to five words or phrases describing what you would see, hear, taste, touch, and smell while imagining yourself enjoying your purchase.

	Sight	Sound	Touch	Taste	Smell
1					
2					
3					
4					
5					

Describe your vision to your partner using your sensory words and phrases as though you have won and are experiencing it in the now. Close your eyes and have your partner repeat your vision back to you in your own sensory words or phrases. For example:

Touch: Feeling of the ocean waves and wind. Feeling of boat rocking.
Sound: Sounds of the ocean or family talking.
Sight: Seeing the sun or moon reflecting over the ocean. Seeing the stars.
Taste: Tasting the sea air. Tasting food or a good meal with family and friends.
Smells: Smelling the sea air. Smelling dinner cooking.

3. Tune into the feeling tone of what you are visualizing. Let the visuals recede into the background and tune into the felt sense of your vision in your body. What is the felt sense in your body right now?

    Describe the feeling tone to your partner. You may find that it's not rational and it's OK to hum or to move your body to communicate your felt sense of your visualization and its feeling tone.

4. With your partner, determine the value(s) you would receive from what you are visualizing. A visualization can represent different values to different people. You must take the time to reflect and define your values.

    For the person who wanted a yacht, we would ask, "What are the values you experience when you have gotten your yacht?" Answers may be:
    - Generosity
    - Family
    - Friendships
    - Adventure
    - Appreciation of nature

    Someone else's values from the same yacht purchase may be:
    - Independence
    - Community
    - Achievement
    - Security
    - Sharing

5. Diverse Value Outcome: It's always good to allow yourself to be open to your values being expressed in many creative forms by stating: I am open to the possibility of my values appearing in many different forms.

6. Choose a Carrier Sound: For "lottery winning," a good universal carrier sound is C & G tuning forks. You could also use a recording of ocean waves, crystal bowls, etc. It's always a matter of determining what the best sound is that supports the listener and allows them to relax. The sound can then bypass defenses and carry the visualization feeling tone directly to the core brain.

    a. Tune into the feeling tone of your vision and allow your body to vibrate with it. Tap C & G tuning forks or play a sound healing instrument of your choice. While doing this, allow your body to move in different ways with the sound.

b. Sit with your partner in a comfortable place. Your partner can take you on a guided visualization of your yacht and your values using a sensory description. For example:

Imagine that you're on your new yacht and you're cruising the Caribbean with your family and friends. You can smell the ocean and feel the breeze blowing a gentle salt spray onto your face. The boat is slightly rocking and you can hear the sound of your friends talking the background. Listen. Tap C & G tuning forks or play sound healing instrument of your choice.

Guided Wellness Visualizations

Guided Wellness Visualizations can be read to yourself, a friend or family member, a patient, or a group. Before you begin reading them and playing sounds, be sure to follow the instructions below.

1. Make sure the person or group you are reading to is safe.
2. Make sure the person or group you are reading to is comfortable. The guided visualizations can be read with the listeners standing, sitting, or laying down. It is most important that they are comfortable and relaxed.
3. The guided visualizations should be read in a relaxed, clear, flowing, rhythmic voice with pauses of different lengths. I suggest you listen to my free trance downloads listed in Appendix F if you are not sure how to read and/or would like to listen to examples.
4. I have suggested tuning forks to use with each of the guided wellness visualizations but voice, crystal bowls, gongs, metal bowls, different musics, bells, etc., can also be used.

DREAM DOLPHINS   For General Wellness

Imagine that you are floating on a calm ocean on a bright sunny day. You naturally relax as your body rises and falls on the gentle waves. There is just the right amount of tension to be able to hear the sounds of the water lapping against your ears. You begin to sense two presences swimming in the ocean underneath you. Deep inside, you know the feeling of these presences and that they have been with you for a very long time. At that moment, they begin slowly swimming to you in a crisscross pattern.

To your surprise, a dolphin surfaces near your left ear and another surfaces near your right ear. As you rise and fall with the waves, they speak to you in telepathic dolphin language.

They say, "We are your dream dolphins. We have been with you your whole life. You can ask us anything you want. You can ask us any questions you want about your life and we will swim the great ocean and will bring back knowledge that is just right for you."

They come even closer and begin whispering. Listen.

Tap C & G tuning forks or play sound healing instrument of your choice.

FOREST PATHWAYS   For General Wellness

Imagine that you are standing on the edge of a lush green forest. An intelligent voice inside of you whispers, "The path that is best for you to take will appear. Take the first step and then take another. Deep in the forest a special secret awaits you. Listen to the signs."

As you take your first step into the forest, a path appears. With each step, leaves and pine needles crunch under your feet. The path meanders through the tall trees. Sunlight appears and disappears through thousands of leaves and branches. The forest is cool and the fragrance of pine and earth soothes your nerves. Your thoughts become slower as you surrender to the rhythm of your walk.

Tap C & G tuning forks or play sound healing instrument of your choice.

Walking down a hill near a small cliff, you come to a pond. The surface of the pond is very still and you can see that just beyond it there are three paths, one on the left, one in the middle, and another on the right. Each path has something special to offer. The question is, "Which path will you take?"

As you look closer, you notice that an owl is sitting on the limb of a tree above the left path. His eyes are big and for a moment he looks into you.

A young doe is grazing along the middle path. She looks up for a moment. Her eyes are a deep dark black.

Along the path to your right, you can see a clearing in the distance filled with sunbeams.

Take a moment and allow your creative self to choose which path is best for you. Each path leads to success. Each path is a new way of understanding and organizing your life experiences in ways you have not yet imagined. The choice lies deep inside of you. Listen and allow your intuitive self to make your choice.

Tap C & G tuning forks or play sound healing instrument of your choice.

As you walk down your chosen forest path, you begin to put together new ideas. You discover and integrate new possibilities. And you are pleasantly surprised at your ability to make new understandings and learnings in order to make life changes that are in alignment with your values. Continue down your path for a while. Relax and enjoy your new learnings.

Tap C & G tuning forks or play sound healing instrument of your choice.

KALEIDOSCOPE   For General Wellness

Imagine that you are in a cool fine mist. A light shines through the mist and each drop of water reflects thousands of rainbow colors. Every thought you have creates a kaleidoscope of changing geometric patterns within the mist. There are patterns within patterns, always changing with each thought. You are deeply relaxed. Allow your creative self to focus on and to learn from your most valuable thoughts.

Tap C & G tuning forks or play sound healing instrument of your choice.

MOUNTAIN TEMPLE   For General Wellness

Imagine that you are walking up a winding mountain path. The air is cool and fresh and the sun is warm on your shoulders. You can hear the sound of your footsteps, one step following another as your breath relaxes into an easy rhythm. Through a break in the greenery, you can see the pathway winding up from below and the landscape stretching out to a distant lake.

Your attention returns to the upward rhythm of your winding climb, the path curving slightly left and then right. You continue, walking further up the side of the mountain. The sun has become hidden by clouds and the temperature drops slightly. Before you reach the summit, the path takes a sharp turn to the right and the sun suddenly blazes forth, illuminating a white temple above you. This temple has six white pillars facing the path and eleven overlooking a cliff. You walk towards the temple, feeling the increased warmth of the sun as it reaches its zenith. As you get closer, you become more excited. When you reach the temple entrance, you take off your shoes, and step up the marble steps to the pleasantly cool floor.

Within the temple, there is row of four pillars. You can see across the floor to another row of four near the far end of the temple. Slowly you move towards one of the closer pillars. Something about this one attracts you.

As you reach this pillar, you know it is time to sit down and so you do, sitting cross-legged with your back resting comfortably against the pillar. You feel the coolness of the floor drifting pleasantly up through your body.

You relax and exhale. You look across the temple floor and see a very old woman dressed all in white. Her face is wrinkled with a deep wisdom. She knows something of utmost importance. Somehow you know her and she seems to be inside of you. She is calm and the silence of the temple seems to encompass her.

She gestures to you and effortlessly raises two bells. It is time to close your eyes and listen.

Tap C & G tuning forks or play sound healing instrument of your choice.

WHITE CLOUD   For Building Confidence

Imagine that you have access to a soft, white cloud that holds a lot of information that you want to keep, even though you may never need it again. When you hear the sound of the tuning forks, take all of your thoughts, beliefs, and reasons that have contributed to your lack of confidence, and transfer them to this special cloud. If any lack of confidence, thoughts, or feelings come to your conscious mind, just let them drift on the sound into your cloud. These thoughts and feelings are all part of an old program that needs to be updated.

Tap C & G tuning forks or play sound healing instrument of your choice.

As your brain clears, a new space will open for you to express your values clearly and confidently. As your special cloud drifts on the winds of life, the sun shines through and illuminates your new confidence. Let each sunbeam, filled with the vibration of your values, penetrate every cell of your body. Remember, your cloud will always pass in front of the sun, but the sun will continue to shine.

Tap C & G tuning forks or play sound healing instrument of your choice.

FLOATING LEAVES   For Solving Problems

Imagine that your problems are like leaves floating down a stream towards a waterfall. Look at the leaves and let yourself be intuitively attracted to one of them. Pick the one that is most important for you to become aware of. Notice that it is just one of many others that are floating together down a stream and that they will all eventually disappear over the waterfall.

Tap C & G tuning forks or play sound healing instrument of your choice.

There is something about this leaf that is very important. Soon it will be gone but there is much you can learn from it. Focus your awareness on the leaf as it meanders, spirals, and floats towards the waterfall. Allow yourself to enter into a new relationship with the leaf. Be with the leaf from a new perspective and learn to appreciate it in a new way. Allow your creative self to put together all your thoughts, to see new possibilities, and to make new discoveries, knowing that eventually the leaf will disappear.

Tap C & G tuning forks or play sound healing instrument of your choice.

MAGIC WELL   For Solving Problems

Imagine that you have discovered a well in a secret dream forest and that it is filled with enchanted water. The bucket next to the well is made of oak and the handle is carved in twisting vines. A gentle breeze moves through the forest and the sounds of the fluttering leaves whisper in your ears. Sunbeams shine between the leaves, creating ever-changing geometric patterns of light. You pick up the bucket and slowly lower it into the magic well.

Tap C & G tuning forks or play sound healing instrument of your choice.

You bring the bucket back up. It is filled with a special water that can create new possibilities, insights, and life directions that have been waiting patiently to come into your consciousness. All your problems, concerns, and worries will dissolve into the taste of the water and return to an ancient place deep inside yourself. It can then transform into well water which is full of new life designs.

The sun is at its zenith, shining warmly through the forest canopy. The bucket is in your hands and it is time to drink this cool and refreshing water. Bring the bucket to your lips and listen as the ancient water flows into your whole self, nourishing new exciting life possibilities.

Tap C & G tuning forks or play sound healing instrument of your choice.

SUNSET   For Pain Management

You are sitting on a beach at sunset with a warm, calm breeze passing over your body. The sun is gradually sinking into the ocean. The sky changes from a glowing bright red to a deep dark red and then becomes a soft orange. As the sun disappears behind the horizon, purple, blue, and violet colors paint the horizon. At that moment, the breeze subsides and a special stillness comes over you.

Tap C & G tuning forks or play sound healing instrument of your choice.

VARIATION I: Enjoy the feeling of quiet tranquility. Take a deep breath. Realize that this tranquility is available to you whenever you need it. As you listen to the sounds, allow them to bring stillness to any place where you are experiencing discomfort or pain.

VARIATION II: Enjoy the feeling of quiet tranquility. Take a deep breath and place your left or your right hand on a place where you feel pain. Invite the source of the pain to come into your inner stillness. Let the source of the pain know that it is an important part of you and that you want to learn how to be with it in a new way. Ask the source to find ways to express itself other than through pain. Tell the source that you want to find a safe way to listen to what it is telling you.

Tap C & G tuning forks or play sound healing instrument of your choice.

DOLPHIN SWIMMING   For Pain Management

Imagine you are a dolphin swimming effortlessly through currents just below the surface of the ocean. Feel the water streaming past you, sometimes warmer and sometimes cooler. Any discomfort you may have had is fading away as you enjoy coasting along the ocean currents. The whole ocean is yours. It is filled with new possibilities and new experiences. Soon you will become the dolphin you are imagining, swimming through an inner ocean of new possibilities. Listen as your human thoughts become dolphin sounds.

Tap C & G tuning forks or play sound healing instrument of your choice.

Any previous discomfort you may have felt is fading away now as you dive deeper into the ocean. The water is soothing as it rushes past you. Feel the cooler currents as you dive deeper and the pressure of the water above you as you adjust to the depths. Become aware of the forests of sea plants and the canyons of coral. There is a moment when you discover a stillness in the currents. In this place, allow your dolphin body to rest. Enjoy as the water slowly drifts to your human self within a vortex of stillness.

Tap C & G tuning forks or play sound healing instrument of your choice.

Take a moment to allow your refreshed human eyes to open. Be prepared to see and experience your human world in a whole new way. Make that special connection with your dolphin self that allows you to dissolve your pain into the ocean currents as your life flows in new and unexpected ways. Listen as your many dolphin friends share their secrets in their special language that you can translate into human thought in your dreams tonight.

Tap C & G tuning forks or play sound healing instrument of your choice.

EMERALD SEAHORSE   For Healing Past Memories of Pain and Hurt

Listen as a seahorse with emerald eyes and a beautiful spiral tail whispers into your inner ear, "I am your dream seahorse. I swim the great ocean of cerebrospinal fluid and bring back only those memories that are important for you right now so that your creative self can take them apart and put them back together in new ways. You can safely learn from your past pain and hurts and leave them behind, knowing they will become nourishment for magical rainbow fishes."

Listen as your dream seahorse whispers and mysterious dreams of mermaids sing in secret melodies as memories appear and disappear.

Tap C & G tuning forks or play sound healing instrument of your choice.

MOUNTAIN PATH   For Healing Tension Headaches

Imagine that you are walking along a high mountain path on a cold winter's day. Step-by-step the tension in your body dissolves as you leave your everyday thoughts behind with each step. Taking a deep breath of fresh mountain air, your eyes relax as you enjoy the mountain views. With each step, a refreshing cold wind touches your forehead and your temples. It is as though someone were holding an ice cube against your temples as a calm, cool, relaxed, tingly numbness spreads over your face and the back of your head. As you listen to the sounds, touch a place on your body where you feel tension or pain until you feel a relaxed numbness.

Tap C & G tuning forks or play sound healing instrument of your choice.

SLOW EATING For Ideal Weight

See yourself sitting down to eat, taking a deep breath, and being very quiet and still. You are relaxed, observant, and your attention is focused on the food you are about to eat. Imagine starting with small bites and chewing them very slowly. Take your time to slowly experience the texture and taste of the food and be pleasantly surprised by what you discover with each bite. Continue eating and learning something new with each bite, something that is just right for your body.

Tap C & G tuning forks or play sound healing instrument of your choice.

Imagine tasting food in new and unexpected ways. Enjoy the aromas of your food. Enjoy the colors of your food. As you continue eating, become aware of an ever-increasing sensation of fullness in your stomach. Imagine knowing with a feeling of certainty, fulfillment, and satisfaction in every cell in your body that you are full and it's time to stop eating. Memorize that comfortable satisfied feeling as you listen to the sounds as it vibrates to the core of your brain and inspires new creative eating patterns.

Tap C & G tuning forks or play sound healing instrument of your choice.

LONG TIDE For Relaxation and Anxiety

Imagine inhaling through your fingertips, up your arms, and into your shoulders. Pause and then exhale down your body into your abdomen and legs and out your toes. Repeat and feel how this slow, deep breathing moves like a gently rolling wave throughout your whole body. Listen to the sounds as you continue to take long, slow breaths. Feel how your energy flows during inhalation and tensions stream out through exhalations. Gradually allow your awareness to be with the sounds and let your whole body naturally breathe in and out.

Tap C & G tuning forks or play sound healing instrument of your choice.

SKIN BREATHING For Relaxation and Anxiety

Scan your body and find a place that is wanting your attention. Bring your awareness to the body surface of that area and imagine inhaling and exhaling through your skin. With each inhalation, feel something special and vibrant coming in. With each exhalation, relax to just the right amount of tension and all allow all the excess tension to come out. Continue to inhale and exhale through your skin and feel how vibrant and refreshed you feel as you listen to the sounds.

Tap C & G tuning forks or play sound healing instrument of your choice.

OK TO SLEEP   For Sleeping

When you listen to the sounds, allow yourself to remember a time when you slept very well. Allow your body to remember that time and learn from it. Let yourself know that it's OK to sleep, and that you will discover new ways to express your new energy when you wake up refreshed.

Tap C & G tuning forks or play sound healing instrument of your choice.

Still Point Interlude

In a world of goals and accomplishments, still points are oftentimes overlooked and undervalued. We cannot see stillness or possess it as a commodity, so it appears to be of little practical use. Therefore, we take moments of stillness for granted or even ignore them as they pass quickly into the routine of daily activity. However, the good news is that stillness is everywhere, it is free, and that we never have to look far in order to find it.

Meditate on the glass-like surface of a still pond. Toss a rock into the pond and watch as the ripples move outwards and return, creating different patterns. Hold the picture of the stillness in your mind and simultaneously enjoy the patterns. Watch the patterns dissolve back into the stillness.

Listen to a wind chime sounding in a breeze. When the breeze becomes still, the sounds of the chimes stop. If you take in the larger picture, you may notice that the trees and plants are still moving but the chimes have stopped sounding. The chimes are momentarily in a vortex of stillness. Listen to the silence the way you listened to the chimes. When the breeze resumes, listen to the chimes as though they were sounding on an ocean of silence.

SECTION FOUR

Sound Musings

The Perfect Fifth:
The Science and Alchemy of Sound**

This paper is based on a hypothesis that explores the ancient archetype of the Perfect Fifth, a sonic interval, and its potential importance in the applications of sound healing in modern stress science. An interval in sound is a precise space between two tones. The Perfect Fifth is a precise tonal relationship defined by a 2/3 ratio that was believed in ancient cultures to have profound healing qualities. The Perfect Fifth is also an archetype that repeats itself over and over to create a vibrational field that gives rise to everything we know. Its healing qualities, well known in the ancient times, will be presented in case histories and in a review of research in modern biochemistry and neuroscience that makes the case for the healing power of sound, strongly suggesting the need for more research.

The purpose of this paper is to learn from and be inspired by the great teachers of the past and to b etter understand their way of conceptualizing the universe and healing in the light of modern science. The paper is divided into three parts. The first part presents an understanding of the interval of a Perfect Fifth and its underlying ancient sound healing principles. The second part presents "The Alchemy of the Perfect Fifth." The third part presents "The Perfect Fifth and Sound Healing" and introduces case histories and scientific insights on the mechanisms by which these ancient sound healing practices work. The intention is to show similarities between the teachings of the past and present day practices based on modern research.

**This article was first published in the 2017 *Rose+Croix Journal*.

INTRODUCTION

One must keep in mind that the great teachings and practices of the past are often expressed in metaphors and stories that have been passed down over hundreds and even thousands of years and that are very different from our modem scientific language. Often one cannot be sure who the authors are, their exact time of publication, or even if their stories have been changed during the course of history. This paper does not pretend to know "the truth" of the past. Instead, it examines literature from the past and asks how it might be understood in the light of modern evidence-based healing practices.

For example, Manly P. Hall relates a Pythagorean story in his book, *The Therapeutic Value of Music*. When I first read it, I thought his story was exaggerated. However, after having an experience with sound after the September 11, 2001, attacks in New York City ("9/11"), I realized that the same story could have been told again in this new modern context. In the Manly P. Hall story the person uses a lyre tuned to Pythagorean intervals. In my story I used tuning forks tuned to Pythagorean intervals.

ANCIENT STORY

A demented youth forced his way into the dwelling of a prominent judge who had recently sentenced the boy's father to death for a criminal offense. The frenzied lad, bearing a naked sword, approached the jurist, who was dining with friends, and threatened his life. Among the guests was a Pythagorean student. Reaching over quietly, he plucked a fifth upon a lyre which had been laid aside by a musician who had been entertaining the gathering. At the sound of the fifth, the crazed young man stopped in his tracks and could not move. He was led away as though in a trance.[1]

A MODERN STORY: DOWN REGULATING STRESS IN A STRESSFUL ENVIRONMENT

One week after 9/11, I flew to Switzerland from Newark Airport to teach a sound healing class in Zurich. The airport was empty because people were afraid to fly. When I went through security, the officers stopped me because I had C & G tuning forks in my backpack. I was brought into an office where an armed customs officer had placed my tuning forks on a desk. The situation was very tense.

He looked at me with hard eyes and asked, "What are these?"

I said, "They're tuning forks. I use them with my patients and I have a brochure in my bag explaining them."

He said, "What do they do?"

I asked, "Can I play them for you?"

He nodded and I picked up my tuning forks, tapped them on my knees, reached across the desk and held them to his ears.

The moment he heard them his eyes softened and then closed. He made a "humming" sound in resonance with the tuning forks. The experience took no more than a minute. He opened his eyes. He was clearly more relaxed. His eyes were brighter and exuded softness.

He said in a much more open resonant voice, "I understand."

Then he said, "Bob, get in here!"

Another officer named Bob walked into his office and I tapped the forks for him. Bob had a similar experience.

Then the officer said, "Thanks, doc, and have a good flight."

I said, "You are welcome." As I left, I looked him in the eye and placed the tuning forks on his desk, put my hands on them for a moment as if to say they are my gift to you. I knew he was not supposed to take them, so it had to be as though I forgot to take them with me. He nodded and I went on my way.

This story is one of many I have experienced using sound based on Pythagorean tuning as part of my practice as a psychotherapist and naturopathic physician. It is the experience learned in these stories that continues to motivate me to better understand sound healing in the context of modern science and evidence based clinical practice while at the same time honoring the traditions of those healers that came before me.

THE PERFECT FIFTH

The mathematical discovery of the Perfect Fifth as an archetype based on mathematics is credited to Pythagoras, the ancient Greek philosopher and mathematician.[2] Pythagoras used an instrument called a monochord to demonstrate the relationship between sound and numbers. A monochord is a musical instrument consisting of a resonant chamber in the shape of a rectangular box and a string stretched across the box.[3]

```
        c
   ⬳⬳⬳⬳⬳
   |   1   |
```

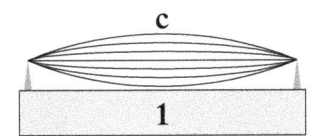

The unsounded string of the monochord represents the potential of creation. The plucking of the string represents the beginning of vibrational creation. Pythagoras assigned the first plucking of the string the number 1. In acoustics, the first sound is called "the fundamental tone." The plucking of the whole string resonates with the archetype of the primordial sound of the birth of the universe, often referred to as "the Word" or "the Logos." From a modern physics perspective the simple act of plucking the string is a metaphor for the "Big Bang" that began our vibrational universe.

When the monochord string is divided into two equal parts by pressing at exactly the halfway point, it creates a sound that is double the vibration of the fundamental or a ratio of 1/2. The ancient Greeks called this sound *diapason*, which meant "through all." In modern music, the diapason is called an octave, which means through eight tones. For example, an octave begins with C256 cps, goes through D, E, F, G, A, B, and ends with C512 cps, an octave higher.[4]

```
      C       C
   ⬳⬳⬳⬳⬳
   |    1/2    |
```

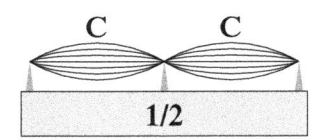

The two tones are the same but double in frequency. They sound the Alpha and Omega archetype, which is expressed in the Hermetic Axiom, "As Above Then So Below." Continuing to divide the monochord string into halves represented the universe dividing itself over and over to create the space for vibrational universes within vibrational universes, all of which resonate with the primordial Logos. In modern systems theory, all systems begin with a Supra System that contains systems within systems that are all congruent with the Supra System.[5]

When Pythagoras divided the monochord string into three equal parts he created a 2/3 ratio and discovered a sound that divided and balanced the octave. He called this sound *diapente*, which meant "through five tones."[6,7] Today, we call the same sound a Perfect Fifth because it is the space between five musical tones. For example C256 and G384 create a Perfect Fifth ratio of 2/3 which spans the five tones of C, D, E, F, and G.[8]

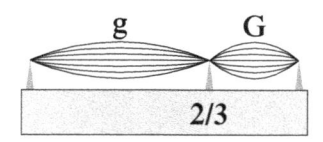

In yoga, the Perfect Fifth is the divine dance between Shiva and Shakti.[9] In Greek astrology, the Perfect Fifth is the light of the Sun that illuminates the whole cosmos. The Chinese philosopher, Lao Tzu referred to the Perfect Fifth as the sound of universal harmony between the forces of yin and yang represented by the image of the Tao.[10]

The Tao gives birth to One.
One gives birth to Two.
Two gives birth to Three.
Three (the Perfect Fifth) gives birth to all things.
— Lao Tzu, *Tao Te Ching*

To better understand the hypothesis that the Perfect Fifth gives birth to all things, one can imagine continuing to divide the monochord string by 3. Doing this will create an ascending spiral of Perfect Fifths.

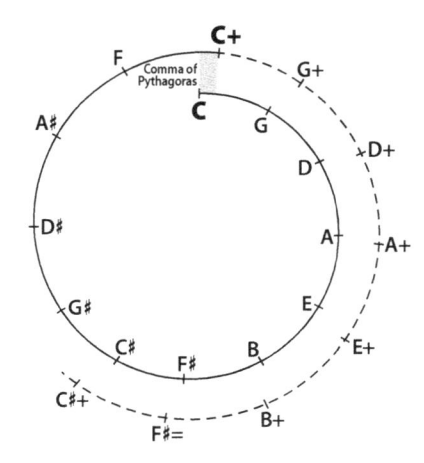

Starting with the Perfect Fifth C–G, the next Perfect Fifth would be G–D, and the next would be D–A, and the next –E, and so on until one arrives at another C that is seven octaves and twelve tones above the starting C. If the twelve tones that appear within the seven-octave span were reduced in octaves into one octave, they would create a twelve tone chromatic scale:

C, Db, D, E, Eb, F, Gb, G, Ab, A, Bb, B, C

However, the second C+ in the diagram, indicated in boldface, will be slightly out of tune with the starting C. In the diagram, the difference between the two C's creates a microtonal interval called the comma of Pythagoras. The comma of Pythagoras can be visualized as a vortex around which Perfect Fifths infinitely spiral. It can also be imagined as a vibrant still point of soundless sound that creates a vortex or neutral center through which all tones of the spiral can be accessed.

"...at the still point, there the dance is...neither arrest nor movement..."[11]

— T. S. Eliot

It is important to think of the Pythagorean spiral as creating a unified field of sound organized around a vortex. To create the unified field, all tones are simultaneously sounding and the continuous tonal spiral is constantly creating microtonal shifts, leading to an oceanic field of waves within waves. In this sense, the Pythagorean tonal spiral is similar to the quantum field.

The division of the Pythagorean tonal spiral by seven octaves and twelve tones represents a complete cycle of ascension. The new C, seven octaves higher than the beginning C, is a slightly different tone that represents a completely new tuned cycle of ascension. Although the Pythagorean comma difference of 1.013643 cps may seem small and insignificant, it is very significant. If one turns a radio dial just a little bit, he/she can tune into a completely different station.

The continuing Pythagorean spiral of Perfect Fifths can also be visualized as a stairway to heaven. William Blake in his famous painting "Stairway to Heaven" shows a spiral staircase leading to heaven.

William Blake's "Jacob's Ladder" (1800) is continual spiral ascending from Earth into higher and higher realms.

Jacob's Ladder, 1805, British Museum, London

THE ALCHEMY OF THE PERFECT FIFTH

The Mundane Monochord by Robert Fludd, often referred to as the World Monochord, is a graphic summary of Pythagorean Universal Sound principles based on a Perfect Fifth, principles that are important to this day. The World Monochord illustrates Pythagorean harmonic principles mapped to elements, planets, angelic kingdoms, and the hand of God.

Pythagoras believed in the Hermetic Axiom: as above, then so below. Looking at the World Monochord, one notices the monochord string stretching between heaven and earth. The string is anchored to the earth below and tuned by the hand of God above. When it is tuned correctly, the monochord string sounds the primordial Logos from which a vibrational universe manifests in octaves and Perfect Fifths. The World Monochord drawing covers a two-octave span; however, it represents an infinite spiral of ascending octaves and Perfect Fifths.

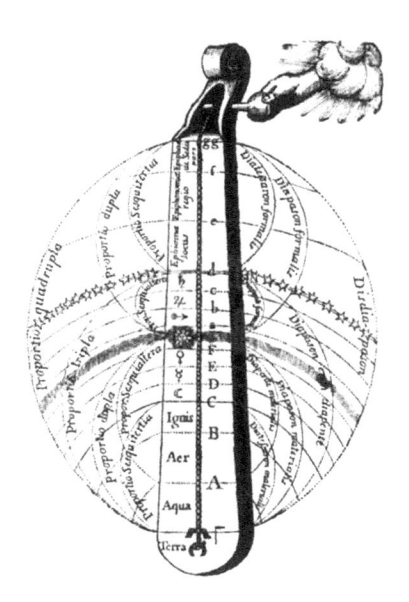

Robert Fludd, "The Mundane Monochord" with Its Proportions and Intervals from *De Musica Mundana* 1618

At the bottom of the monochord string are the four primordial elements. Fludd presents them in ascending order Earth (Terra), Water (Aqua), Air (Acr), and Fire (Ignis). The four elements are bound together into a four-note tetrachord. A tetrachord is an interval of a Perfect Fourth that contains within it four notes that consist of two whole tones and a half tone.

Tetrachord

Each note is ½ step, i.e., C to C# is ½ step. C to D is a whole step or whole tone. In the diagram, the four notes are C, D, E, and F. In order of steps, they are whole step (C-D), whole step (D-E), and half step (E-F).

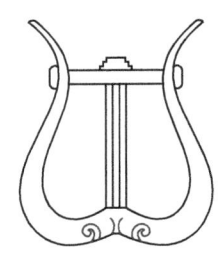

It is said that the first lyre of Hermes had only four strings and, when played correctly, the four elements came into balance allowing the listener to ascend though the planets into the Empyrean realms of heaven.[12]

The last element in Fludd's primordial tetrachord is fire. When the four elements are balanced, the primordial fire burns the earth and creates alchemic heat. This is called the fire of alchemical transmutation through which one is able to ascend through the planets into the Empyrean realms. This is often represented in alchemical drawings as the phoenix rising from the ashes.

"Melchizedek and the Mystery of Fire," 1996,
the Philosophical Research Society, Inc., Los Angeles, CA[13]

In the Pythagorean tonal cosmos, the octave consists of eight tones that are divided into lower and upper tetrachords.[14]

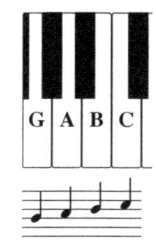

Lower Tetrachord

Upper Tetrachord

The two tetrachords can be visualized as a sonic archetype that represents a higher–lower self-relationship within oneself. The lower tetrachord contains the four elements of the lower self and the upper tetrachord contains the same four elements transformed into the elements of the higher self. The elements of the lower self are associated with the challenges and struggles of life. In modern science they are associated with distress patterns. The elements of the higher self are associated with higher states of consciousness, light beings, and wisdom. In modern science, they are associated with eustress and peak experiences. When the elements of the lower self are balanced, one naturally reaches the interval of a Perfect Fifth, which is the perfect balance between heaven and earth.

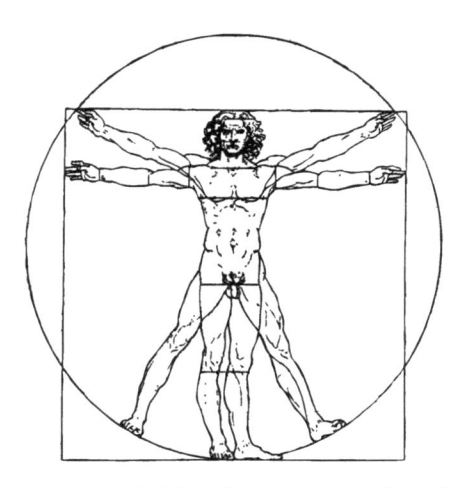

Leonardo da Vinci "Squaring the Circle"

Leonardo da Vinci's drawing "Squaring the Circle" illustrates the principle of the Perfect Fifth balancing the higher and lower self. The square represents the four lower self elements; the circle represents heaven and the four higher self elements.

The Perfect Fifth is graphically represented by the Vitruvian Man simultaneously touching the square and circle.[15]

The sonic relationship between the tetrachords and the element archetypes has been used for thousands of years in different knowledge systems to demonstrate a transformation of lower self elements to higher self elements. In Western alchemy, the higher–lower self-relationship between the archetypal elements was represented as bottles within which the elements mixed in different patterns. The flasks were called the quintessence, which means the essence of five that sounds a perfect balance of the four elements.[16]

The Manly P. Hall collection of alchemical manuscripts[17]

The two quintessence flasks should be imagined as one flask that contains different element patterns that represent a higher–lower self process of transformation that takes place within. The alchemists often referred to this process as the transmutation of lead, the lower self, to gold, the higher self.

For purposes of illustration, the flasks look the same; however, in life the space within the flasks is constantly changing based on the amount of space necessary for the four elements to properly mix. In daily life, one is always needing more or less space. If the space is too small, one feels confined, trapped, stuck, or compressed. If the space is too large, one feels scattered, spaced out, and lost. If the space is "just right," one gains energy and has the space to pull together his/her resources to resolve life challenges. When one has just the right amount of space, it is called a "tuned space."

The element patterns within the flask can be sonically imagined as tones that are playing louder and softer in different orders. Within the first flask, the elements entangle in a lower self-distress pattern. The dove represents one's consciousness being drawn into the pattern in order to resolve the distress. The alchemists referred to the element pattern in the first flask as Prima Materia. Today we call

Prima Materia distress. For the alchemists, Prima Materia was "the stuff of our lives" and among the numerous terms used to describe it were feces, urine, and saliva. Today, "shit happens," an all too familiar slang expression spoken during experiences of distress, captures this sentiment well. The goal of an alchemical operation is to transform Prima Materia, the lower self, into the light, or higher self. Today, one seeks to transform distress into eustress.

One's consciousness is continually descending into distress, resolving distress, and ascending into eustress. All element patterns are good patterns, and when one learns how to navigate an element pattern, he/she becomes a master of the positive energy that the pattern offers. Imagine a hang glider soaring like a dove in the winds. If the winds change, the hang glider simultaneously adapts to the changing pattern. If the hang glider fails to adapt, he/she will experience increasing levels of distress leading to a crisis. If one senses the wind change, he/she will creatively adapt to the new wind pattern. The more wind patterns the hang glider learns how to navigate, the greater his/her mastery and euphoria. The ancient sages of India referred to constantly changing element patterns as Shabda, which means the currents of Sacred Sound. These currents flow within a universe of constant energetic change. The consciousness of enlightenment requires continuing adaptation and change to these currents.

The ascension of the dove shown in the higher self quintessence flask was called "the web of Athena" by the Greeks.

Persephone and Hades, The British Museum

The web of Athena represents increasing euphoric states of consciousness that are made up of higher self elements that also must be balanced. Just as one can get lost in one's lower self distress experiences, one can also get lost in the euphoria of the higher self experiences. This is what is called being caught in the web of Athena. When one is able to navigate the euphoric elements of the higher self tetrachord, one gains mastery of the C–C octave and enters a new octave via a Perfect Fifth. The elements of the higher self are the tetrachord G, A, B, C. When these elements are balanced via a Perfect Fifth, one comes to the beginning of a new octave represented by the note or tone D (G–D is a higher self balancing Perfect Fifth). Each new octave is a new space filled with new elemental challenges.

This sort of vibrational field understanding of reality is challenging to comprehend within a reductionist worldview. The ancient astrological texts used to talk about Venus becoming Mars. This makes no sense if one believes that Venus is a planet and Mars is a planet. It is impossible for one planet to transform into the other planet. However, if the planets are vibrational symbols like notes, then it is possible. In the night sky, Venus and Mars are seen in different places; however, in the inner self Venus can, like tones of a lyre, harmoniously merge with Mars and eventually transform into Mars. Paracelsus, the Swiss alchemist and great-grandfather of modern medicine, understood this process in a different way before the discovery of quantum mechanics.

> The physician should know how to bring about a conjunction between the astral Mars (quantum nonlocal Mars) and the earth Mars (psychological individual Mars). In this sense, the remedy should be prepared in the star (quantum field) and should become a star (within the individual), for the stars above can make us ill and die, or they can make us healthy. As a remedy cannot act without the heavens (quantum field), it must be directed by them.[18]

In the Pythagorean harmonic cosmos, poetically called the Music of the Spheres, the ascension process is one of moving through different octaves of creation that are given planet names. Each octave has its own unique planet qualities and harmonies that the Greeks called modes. The Pythagorean harmonic stairway to heaven ascends through Seven Octaves and 12 tones. What are called sharps and flats in the modern chromatic scale are the five tones that the Greeks called Chroma, which means to give color to. By altering different tones within the seven octaves, they created different tonal colors or moods that correspond to the qualities of each octave. Modes in classical Indian music are called Raga, or "that which colors the mind."

Apollo, God of music and healing, playing modes on his seven stringed lyre. Apollo is from a painted interior of a while ceramic kylix found at Delphi and dated at circa 470 BC.

Modes are the basis of healing music played on the seven stringed lyre of Apollo. Through tuning to and sounding modes in different melodic patterns, the musician healer can guide a person through the web of elements and planets into the stars.

Traditionally, modes have been used to create melodies that have healing powers and that are called Medicine Melodies.[19, 20] Beyond their use in music, modes can be used in sound healing to create sonic fields and medicine melodies with tuning forks. Within these fields, the listener will hear spontaneous melodies appear and disappear. However, what is most important is immersion into a modal field of sound through Mindful Listening. Mindful Listening is similar to Mindfulness Meditation, and the two techniques are complementary and easily combined. Each practice aims to go beyond the "rational-objective mind" to expand consciousness and to increase one's awareness. During Mindful Listening, practitioners listen to a sound and become one with it.

The following chart gives the formulas for creating modal sonic fields from BioSonic tuning forks and the healing qualities ascribed by the Greeks to each field.[21] They have been recorded and are available for listening on my CD *Apollo's Lyre: A Modern Adventure in Ancient Cosmic Harmony.*[22] The relationship between the modes and planet archetypes have been discussed and debated by many different authors over thousands of years and are given as generic historical references which are not intended to be factual descriptions.[23] The modal recordings are presented as listening experiences based on the creation of the modes from the Perfect Fifth. The individual tones are sounded as a part of a whole field. For example, the only difference between the Ionian Mode and the Lydian Mode is that F becomes F#. However, a change in frequency, no matter how small, has the ability to affect the entire field and the consciousness of the person entrained within that field. In this case, the change from F to F# creates a whole new field with different qualities.

BIOSONIC MEDICINE MELODY TUNING FORK TONES USED IN APOLLO'S LYRE RECORDING[24]

Locrian (Moon)	C256	Db	Eb	F	Gb	Ab	Bb
To calm down after a stressful day and to promote good sleep and dreaming							

Ionian (Mercury)	C256	D	E	F	G	A	B
To inspire clear communication and putting together thoughts to reflect inner vision							

Aeolian (Venus)	C256	D	Eb	F	G	Ab	Bb
To stimulate artistic creativity through quite dreamy/reflective moods							

Dorian (Sun)	C256	D	Eb	F	G	A	Bb
To awaken and clear the mind—good in the morning. It was the most used mode in Gregorian chant.							

Phrygian (Mars)	C256	Db	Eb	F	G	Ab	Bb
To awaken motivation and drive to reach a goal							

Lydian (Jupiter)	C256	D	E	F#	G	A	B
To promote good feelings, happiness, and a sense of celebration of life							

Mixolydian (Saturn)	C256	D	E	F	G	A	Bb
To support increased focus and clarity							

The medicine melody tones are presented as pure sound, based on a precise 8 cps fundamental tone and interval ratios in order to encourage phenomenological listening. This gives the listener the option of sharing their experience of a mode with others based on an established baseline. It's up to the listener to mindfully enter into the mode and share their experience in much the same way people might have done in ancient times.

The following are questions that each listener can use to help them systematically explore and journal their experiences of modal listening.

1. General Observations/Impressions
2. Specific Observations
   a. Do you experience a color or colors as you listen to a mode?
   b. Do you experience a temperature change as you listen to a mode, i.e., cool, hot, warm, or cold?
   c. Do you receive thoughts when listening to a mode, i.e., messages, insights, stories, visions, instructions?
   d. Do you experience specific emotions when listening to a mode?
   e. Do you experience a shift or change in your body posture when listening to a mode?
3. Additional observations

The planet archetype of the Perfect Fifth is shown in Fludd's World Monochord as the Sun. The Pythagorean universe was geocentric. Looking out from Earth the first tetrachord is Earth, Moon, Mercury, and Venus. The next tetrachord is Sun, Mars, Jupiter, and Saturn.

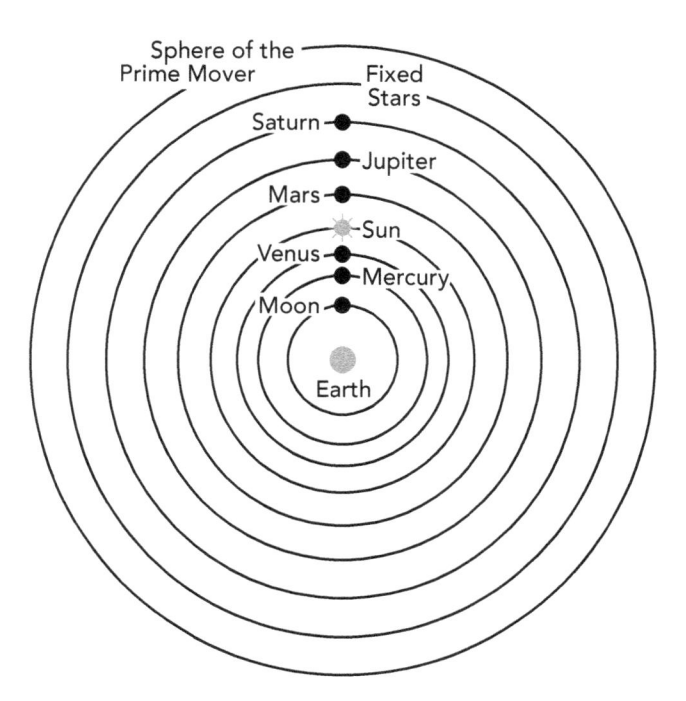

Although it would seem at first glance that the Earth is the center of the universe, it is actually the formation of an interval of a Perfect Fifth by the Earth and Sun that is the center of the universe. Psychologically, everything revolves around our inner sun. The outer Sun in our solar system resonates with one's inner psychological sun that illuminates the whole universe and all its interconnections. The astrological symbol for the Sun is a circle with a dot in the middle. In Eastern mysticism, the same symbol is called Bindu, a point through which the mind transcends into the light of the Absolute. In the East this is called enlightenment or cosmic consciousness.

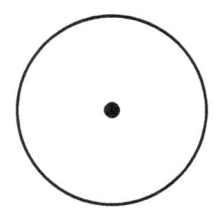

Sun Symbol or Bindu

The astrological symbol for the Earth is a quartered circle where each quarter represents an element much like Fludd's primordial tetrachord where each note represents an element. **By balancing the four elements, we are able to ascend into the inner light of the primordial Sun. We become the archetype of the phoenix rising from the ashes.**

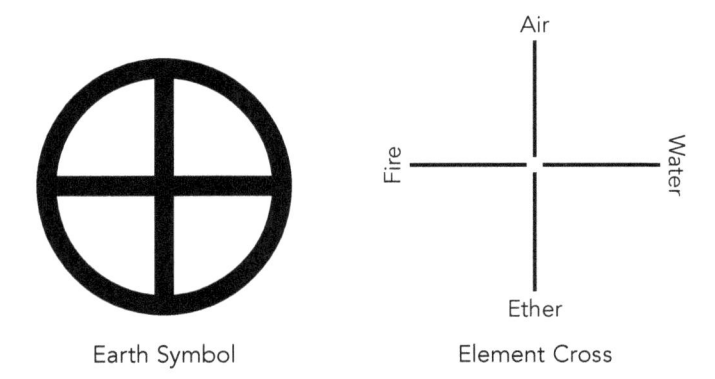

Earth Symbol

Element Cross

THE PERFECT FIFTH AND SOUND HEALING

Sound healing is the practice of using sound in a therapeutic manner to enhance the body's natural ability to heal itself and to promote personal growth and development. The basic premise of sound healing is:

All existence is vibratory in nature and, therefore, it is the underlying vibratory field that sustains and imbues everything that exists with structure and form. The driving force of all healing processes is consciousness, an unexplained fundamental manifestation of the universe that influences the nature and structure of existence by its effect on the behavior of subatomic particles. Sound is a perfect tool for healing practices as well as personal development because it mimics the nature of existence and affects individuals at all levels: anatomically, physiologically, emotionally, psychologically, and spiritually.[25]

There are many sound healing instruments. These include tuning forks, crystal and Himalayan singing bowls, gongs, whistles, didgeridoos, vocal sounds, flutes, and rain sticks. Anything that makes sound, though, is potentially a sound healing instrument. This paper focuses on BioSonic tuning forks that are tuned to a Perfect Fifth.[26]

The primary tuning fork system used in the world today by sound healers are the BioSonic tuning forks.[27] BioSonic tuning forks were developed by the author and include both psychoacoustic and vibroacoustic and forks. The most widely used BioSonic tuning forks by practitioners today are the C & G (Bodytuners™), which are tuned to a Pythagorean Perfect Fifth ratio of 2/3. The Perfect Fifth is part of the Fibonacci series, a mathematical sequence that is part of a broad level of design patterns found in nature, for example, in the human body (extending to the structures of DNA, brain microtubules, and even the pattern of teeth), galactic shapes, the shape of hurricane clouds, breeding patterns of rabbits and bees, the structure of certain chemical compounds, and the spacing of leaves in plants.[28] This pattern is also emulated in architecture and is even reflected in stock market patterns.[29]

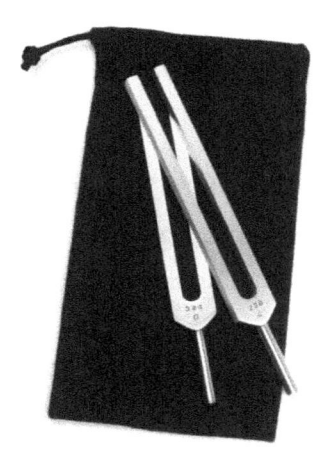

BioSonic C & G Tuning Forks are easy-to-use sound healing instruments that create a Perfect Fifth tone of C & G.

BioSonic tuning forks are versatile sound healing instruments because they are always in tune, are lightweight, can be easily carried in one's pocket, and are easy to learn and use. They lend themselves to research and to consistent results with clients because of their tuning accuracy. Tuning forks are neutral instruments because they are not associated with any specific culture or style of music. The Perfect Fifth tuning forks can be sounded by tapping them together and/or tapping them on the patella bone. Each tap creates a different effect. Tapping them together is louder and rings many overtones. Tapping them on the patella bone is softer and is the method for direct ear use.

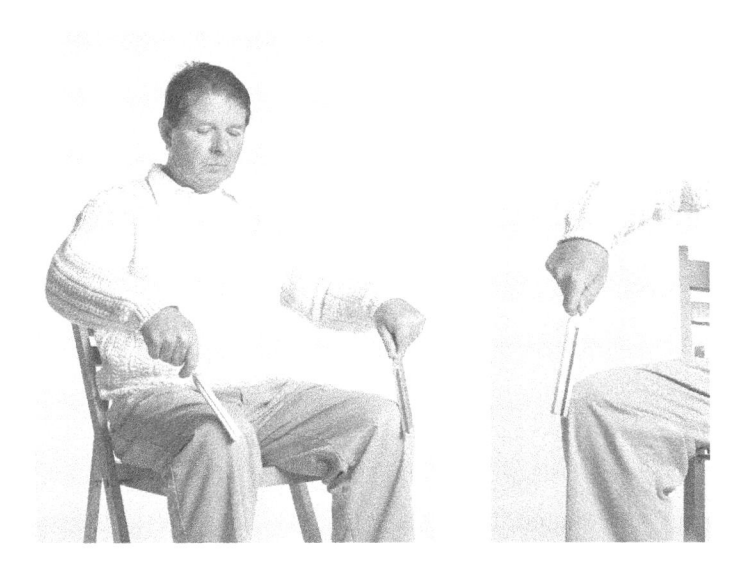

The following case history illustrates the use of the Perfect Fifth and the science behind what happens to the listener.

Case History[30]

Jean (not her real name) was a patient at the Bircher-Bennner Medical Clinic in Zurich, Switzerland. At age 54, she had been diagnosed with advanced metastatic breast cancer. Her prognosis was not good and her oncologist suggested to her that she get her personal affairs in order. The nursing staff requested that I speak with her because after her talk with her oncologist, she had been emotionally acting out and could not sleep.

During our first visit, Jean immediately said in an angry voice, "I don't know what you are going to do for me."

In order to better convey Jean's case and demonstrate the establishment of a therapeutic alliance within a sound healing process, I will tell the story in the first person.

I said, "Jean, I have read your chart and I want to do something for you. And I know you do not know what it is."

She replied in an angry voice, "I do not think anyone can do anything for me. I certainly do not see how talking can help."

I then said to Jean, in a moderately loud and direct staccato voice, similar to her voice, but without the angry tone, "Jean, I know you think that no one can do anything for you. I do not want to talk with you. I want to play some sounds for you. There is nothing you have to do except lie there and listen. It will only take a few minutes and then I will be gone."

An important part of sound healing is to form a therapeutic alliance with the patient. This requires going to the patient's reality construct in order to lead them to new options. In order to play the tuning forks for Jean, I reflected her words back to her in a moderately loud staccato voice. This is a form of voice sound healing called Voice Energetics, which is fundamental to a successful therapeutic alliance.[31]

Jean immediately responded to the sound of my voice and her demeanor and vocal tone changed.

She said in a quiet, more flowing voice, with less of an angry tone, "OK, as long as there is no talking."

At this point in the sound healing process, the therapeutic alliance was formed. Jean was now ready to receive the sound. I took the C & G tuning forks (Perfect Fifth), showed them to her and gave her instructions to close her eyes, be with the sound, and let the sound take her to exactly where she needed to be. She closed her eyes; the tuning forks were tapped with healing intention and brought to her ears. When they stopped ringing, I left her room as promised without saying another word.

That evening the ward nurse came over and asked me, "What did you do to Jean? After you left, she went to sleep and when she woke up she was a different person. She did not complain, and she stopped asking for pain medication. She even talks with me without being angry. She is also asking to see you again."

The next morning I visited Jean. She was very happy to see me. Her voice quality was softer and less staccato and she was not angry. Our therapeutic alliance was established and she felt safe. She took my hand and said, "Would you give me those sounds again?" I sounded the tuning forks for her again and left the room.

Dr. John Beaulieu listening to C & G tuning forks that were sounded using the knee tap method.

The next time I came to see her she asked, "Can we talk?"

The therapeutic alliance, once established, forms a container for the patient to safely express themselves on many levels. Originally, Jean did not want to talk, and now it was important for her to talk. She shared that she was an artist, and that she had been suppressing her creative talents for many years. She wanted to paint again.

I asked, "When I come to see you, will you draw our sessions?" She said, "Yes."

When something important is withheld or suppressed it will naturally seek expression as part of the healing process. Our therapeutic alliance established the ground for safe expression and in Jean's case, the sound healing opened the space for her to express herself. Jean was asked to draw her sessions in order to be congruent with her expression and to reinforce the value creative expression represented to her.

Below are two of Jean's drawings. The first is how she experienced me coming to her bedside. The trombone represents sound and the tuning forks. It is interesting because Jean intuited the stars that represent the sounds. The five star geometric pattern is mathematically created by the C & G tuning forks.[32] She pictured me as a lion because I spoke to her in a loud and staccato voice when we first met.

The next drawing represents Jean's feeling of being in the sound. The sound uplifts her. The darkness underneath represents her cancer. She spreads her wings and is ready to fly. The overtones created by tuning forks were in ancient times called the stairway to heaven. Jean is flying into the Sun on her internal uplifting feeling.

In a hospital, patients and staff talk, and it was not long before patients and staff were asking for "the sounds." Whenever I walked into a patient's room to talk with them, they would ask for "the sounds."

As for Jean, her prognosis was not good, and she was scheduled to be sent home to die. To our surprise, before we could tell her, she decided to leave the hospital, move to Italy and paint. During my last visit with her she said, "If I am going to die, I am going to live as an artist." Seven years later, Jean passed away peacefully in Italy surrounded by her paintings.

This case illustrates perfectly how a practitioner can learn to listen beyond the emotion and content of what a patient is saying to the sound dynamics of their voice. By doing so, the practitioner can establish congruency with their patient on a sensory level and tune into the feeling tone underlying a patient's expression. The patient will experience a sense of being heard and a feeling of bonding with the practitioner. Once the practitioner/patient bonding happens, it forms the basis of a therapeutic alliance. Successful doctors and therapists do this intuitively and it has been given names like "good bedside manner." It can also be learned and refined through Voice Energetics as a form of sound healing.

It is hypothesized that the sound of the tuning forks bypassed Jean's protective mechanisms and transported Jean her to a place inside herself that she had suppressed for many years. Her drawing of upliftment might be explained by the release of anandamide molecules leading to a feeling of euphoria, which then lead to a release of nitric oxide (NO). NO is a signaling molecule known to have antiviral, antibacterial, and antitumor properties.[33] **Within this sense of relaxed euphoria, Jean was able to get in touch with a deep desire to paint and express herself within the safety of the therapeutic alliance. By taking responsibility and acting on her desire, she continued to express a new anticancer physiology, which supported a continuing rhythmic release of nitric oxide (cNO), then giving rise to a state of remission.**

SCIENTIFIC DISCUSSION

Research suggests that when individuals listen to music and/or sounds that are safe and enjoyable, they will experience peripheral vasodilation, warming of the skin, a decrease in heart rate and an overwhelming sense of well-being.[34] In 2003, John Beaulieu and colleagues published a peer reviewed article in *Medical Science Monitor* suggesting the physiological pathways through which sound and music work.[35] Specifically, it discussed how sound and music could have the ability to bypass the limbic system and amygdala and go directly to the core brain resulting in the release of anandamide, an endogenous endocannabinoid, which causes the release of cNO in immune cells, neural tissues, and human vascular endothelial cells.

Representative connections among the limbic-hypothalamic pituitary adrenal axis demonstrate that these centers are linked to vascular tone regulation. This pathway suggests how nitric oxide spiking may exert a level of top-down control of vasomotor activity and circulatory tone. **The positive reaction to nitric oxide wave is reduced blood pressure, lower heart rate, greater pain tolerance, overall lowering of metabolism, and a greater sense of well-being and ability to adapt to stressors.** Sound plays a role in the management of stress and anxiety also through its ability to increase or decrease both cortisol and norepinephrine[36] and by its related ability to decrease arousal due to stress.[37] It affects the immune system by stimulating the production of IgA and NK cells.[38]

When this happens, patients will report an experience of inner warmth and a deep sense of well-being. Psychologically, they will be more positive and better able to cope with their environments, resulting in the continued neutralization of the stressors that were inhibiting natural cNO production in immune, nerve, and endothelial cells. The patient will experience less distracting physical and emotional pain due to the release of endocannabinoids and will be more able to focus and talk about what is most important to them.

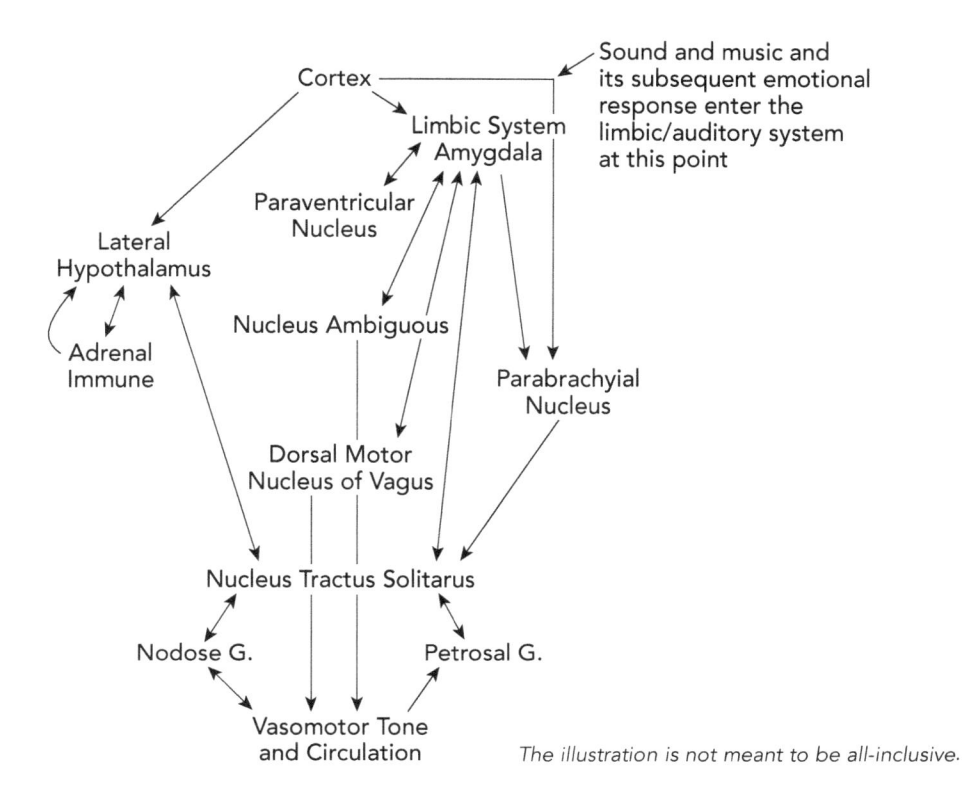

```
                        Cortex ─────────┐    Sound and music and
                           │  ╲          │   its subsequent emotional
                           │   ╲      Limbic System   response enter the
                           │    ╲     Amygdala        limbic/auditory system
                           │     ╲   ▲▲▲              at this point
                           │   Paraventricular ╱
                           ▼     Nucleus  ╱
                     Lateral              ╱
                  Hypothalamus           ╱
                     ▲ ▲ ▲              ╱
                     │ │ │             ╱
                     ▼ │ │   Nucleus Ambiguous
                   Adrenal
                   Immune                   Parabrachyial
                                              Nucleus
                               Dorsal Motor
                              Nucleus of Vagus

                            Nucleus Tractus Solitarus
                             ▲                    ▲
                           ╱                        ╲
                     Nodose G.                    Petrosal G.
                         ▼                            ▲
                            Vasomotor Tone
                            and Circulation        The illustration is not meant to be all-inclusive.
```

In technical terms, nitric oxide is a "gaseous diffusible modulator" that moves through the entire body and central nervous system in waves of gas. The release of nitric oxide counteracts the negative effects of the stress hormone norepinephrine. The presence of norepinephrine results in a racing heart, high blood pressure, anger, anxiety, and greater vulnerability to pain. The positive reaction to the nitric oxide wave is increased neural plasticity, reduced blood pressure, lower heart rate, greater pain tolerance, and overall lowering of metabolism. Psychologically, this leads to less anger, a strong sense of purpose, and a greater sense of well-being, leading to an increased ability to adapt to stressors.

In general, when the tuning forks are tapped with healing intention, their effect is quick and can be integrated with and will enhance every therapy. It is also suggested that the effect of tapping the tuning forks with healing intention will have a similar effect upon the person sounding the tuning forks. Hospitals are stressful and doctors, nurses, therapists, and support staff have a lot to do. Tapping the tuning forks, just for a moment, will stimulate the above physiological and psychological processes. It is a way of shifting gears with a patient and then moving on to the next patient, knowing that the sound will serve to enhance whatever therapies the patient is receiving.

CONCLUSION

Although more research needs to be conducted, I believe that all musical intervals, when mindfully listened too, have the potential to spike nitric oxide. The most important factors in nitric oxide spiking are ease of listening, mindfulness, and the therapeutic relationship. My clinical experience with many patients suggests that the Perfect Fifth is the easiest of all intervals to listen too. I have presented this ease of listening in the context of ancient knowledge with the intention of understanding it in a modern science paradigm. Nitric oxide is a very reactive gas and thus lends itself to the placebo effect[39] which I hypothesize is a natural body response within a healing environment that supports mindful listening. It is within such an environment that the experience of the Perfect Fifth can safely be further researched and explored by all individuals interested in healing.

The physicians of the past worked from a vibrational model of the universe similar to our modern understanding of a vibrational quantum field. They understood that everything was interconnected. Their professional language was one of archetypes and the application of those archetypes to the challenges of everyday life. Comparing their archetypical language to today's modern stress science reveals the depth of their understanding of stress and their ability to work with stress-related diseases. Hans Selye said in his book *The Stress of Life* that his general adaption syndrome could have been discovered in the Middle Ages or earlier through an unbiased state of mind.[40] Although these doctors did not have modern biochemistry or scientific procedures, they were nevertheless doctors who closely observed their patients and learned from their behaviors.

We have much to learn from the physicians and healers that have come before us. They left us with a puzzle that needs to be pieced together by our modern scientific understandings. We may never know exactly how those systems were practiced but we can learn from our experiences inspired by their work and apply them to evidence based science. Ultimately, the clinical outcome, whether it be through phenomenological experience, reductionist science, or systems integration, will always suggest the need for more research, just as the Pythagorean spiral of the Perfect Fifths will always be an infinite continuum.

Disappeared Sound Interlude

nano thin light filaments intertwined into a trillion Gordian knots vibrate as nodes embedded in an undulating oceanic web. The forever secret of the Grail is whispered in unforeseeable patterns, forming and unforming in rainbow whirlpools.

nam sha

nam sha

e

I am the disappeared listening. Eyes upwards sensing the tangerine horizon dissolving. You are my secret within the forever secret that I am telling until we are all secret. The exiting is near oh dream lords of unfathomable sound. The light is vibrant in the non existence of your illusions.

i

pha na

au

I am the vanished leaving on an ever continuing drone created by crystal microtubules resonating the unfathomable sound within my hollow cranium. Oh my special lapis beings with Rose Quartz eyes, If only there were time, enough time, to tell the whole secret.

sa

sssssssing e nam va

e sa   e sa   e sa

Written while listening to Silva Nakkach's CD Liminal

Annotated Nada Bindu Upanishad

The Nada Bindu Upanishad was written around 100 BCE. Nada means sound and bindu is the point from which universal energy originates and flows through all existence. Before the term "quantum" was invented, the sages of the East referred to the universe as Sacred Sound. They used sounds we hear with our ears to enter into the larger experience of the vibrational dynamics of the whole universe. They understood through direct experience that sound mimics the vibrational nature of reality. By "being in sound," they learned about our innermost relationship with the universal field of vibration. Their learning involved living on the deepest levels in resonance with the vibrational nature of reality.

The mind exists as long as there is sound, but with the cessation of sounds, there is the state of being above the mind. The sound is absorbed in the Akshara (indestructible), and the soundless state is the supreme seat. The mind, which along with Prana has its Karmic affinities destroyed by the constant concentration upon Nada, is absorbed in the unstained One. There is no doubt about it. —Nada Bindu Upanishad

The mind exists as long as there is sound

The mind is quantized energy between nodes. In this sense our mind is much more than our everyday thoughts. **It can be conceptualized as a quantum navigational interface though which we experience the infinite vibrational diversity of life.** Our mind interprets quantized waves of energy into patterns and structures that we call reality. Our mind gives us our bearings on an infinite sea of energy, and gives us our location in time and space. Our mind tells us who we are and where we are. There are an infinite number of quantized states of mind. **If we change our tuning, we change everything.**

but with the cessation of sounds,

The cessation of sounds is a when our mind, as quantized waves of energy, enters into a harmonic node. The experience of harmonic nodes are often times called still points by sages, healers, and mystics.

there is the state of being above the mind.
Another way of saying this is that there exists a state of consciousness in which the mind is not engaged. This state of consciousness is the disengagement of the mind, as quantized waves of energy, within a node. The term "above" is best understood as "all pervasive" because the state of being is not above anything. It is infinitely pervasive through everything.

The sound is absorbed in the Akshara (indestructible), and the soundless state is the supreme seat.
When the frequency of a quantized state of mind, and all that is associated with it, i.e., thoughts, emotions, beliefs, biochemistry, physiology, and body structure, enters into (is absorbed in) the stillness of silence (Akshara), we have the ability to shift into and out of all quantized states of mind. We are indestructible whereas all quantized states of mind that we engage are destructible.

The mind, which along with Prana has its Karmic affinities destroyed by the constant concentration upon Nada, is absorbed in the unstained One.
Mind is quantized energy and energy is Prana. Karma is action that creates vibrational effects. When we drive our karmic vehicle (our body) through quantized vibrational realties we create vibrational effects or waves. We create karma through our actions. Karma can be summarized as whatever you reap in life is what you sow.

Through highly focused mindful listening on sound (Nada) the vibrational waves of actions can return to, be absorbed in silent stillness (the unstained One).

There is no doubt about it.

Silence is the language of God. All else is poor translation. —Rumi

Be still, and know that I am God. —Psalm 46:10

In silence the teachings are heard;
In stillness the world is transformed. —Lao Tzu

Quantum Cowboy

Dream Lights in mind space

Neurons without thoughts

Synaptic dreams of chemical canals

Lovers drift in spirals

Sea monsters swim without direction

Quantum Cowboy split my soul

Quantum Cowboy make a hole

through this light beam dreamer

A porthole to see the space saucers disappear

A porthole to let in the nothingness

A porthole to see through

A porthole, a hole in my soul for the light in my eyes

to illuminate the unknown

A porthole to discover myself and disappear my soul

oul

ul

l

Appendices

APPENDIX A
How to Sound BioSonic Tuning Forks

The BioSonic C & G tuning forks, called Body Tuners™, can be sounded by tapping them together and/or tapping them on the patella bone on the knee. Each tap creates a different effect. Tapping them together is louder and rings many overtones. Tapping them on the patella bone is softer and is a method for direct ear use.

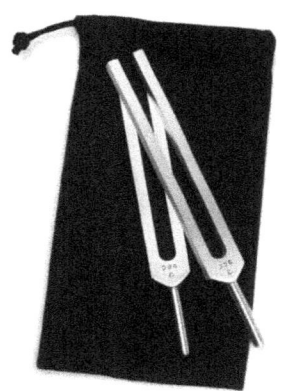

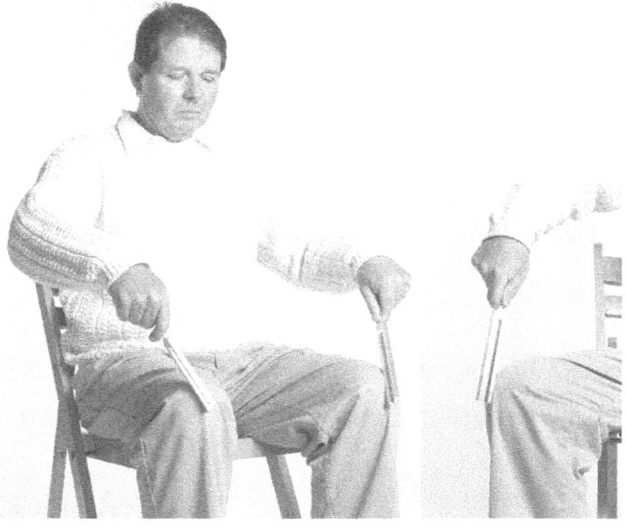

Patella bone tap

Dr. John Beaulieu listening to C & G tuning forks that were sounded using the knee tap method.

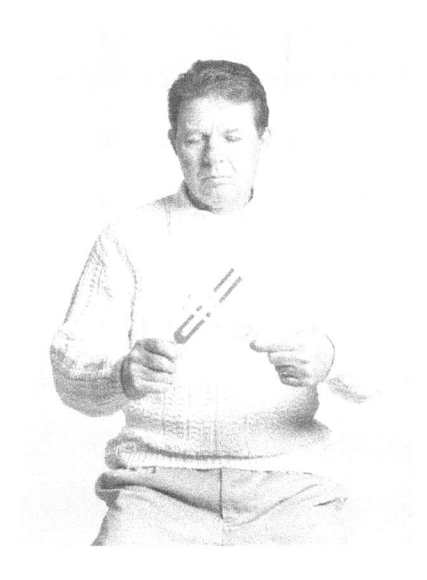

Tap tuning forks together on their edges. Experiment with creating soft taps that gradually get louder and then going from loud taps to soft taps. When you tap tuning forks on their edges, do not bring them to your ears as demonstrated in the knee tap method. Instead, move your body and simultaneously move the tuning forks through the air around your body. Dance with the sounds of the tuning forks.

I suggest that you view my video on how to tap tuning forks in the support section of www.biosonics.com.

APPENDIX B

Human Tune In™ "A BioSonic Sound Healing Concert"

We live in an oscillating universe of infinite possibilities. We are vibrational beings. We resonate. We travel through multi-dimensional entangled filaments of light. We are navigators on a vast ocean of fluidic ambrosia. We hyper-leap through synaptic vortexes surrounded by entangled webs of illuminated light filaments. We are light beings of infinite possibilities dancing in a cosmic womb of stillness to the sounds of a musical universe… —John Beaulieu

Human Tune In™ is performed with precisely tuned BioSonic tuning forks to create sounds for healing, to expand consciousness, and to explore sonic realities. The general healing suggestion given for a Human Tune In™ is to say to yourself:

> I put this experience into the light for the highest value of all who listen.

If you want to be more specific, here are some possibilities you can consider:

> I put this experience into the light and ask that the sound reveal new possibilities for healing any condition. (name a specific condition)

> I put this experience into the light and ask that I, or anyone listening, can discover new ways of solving (life challenges or problems)

> (*Distance Healing*) I put this experience into the light and invite (name person or persons) for healing, knowing that the sound can reach across time and space for them.

LISTENING

A Human Tune In™ concert is different than a normal concert. During this concert, within the limits of the space, you can sit, lie down, and/or move around at any time. Allowing your body to move is very important. In a Human Tune In™ concert, you should be open to allowing your body to move. Ideally, your body would just move. Your body's movements may sometimes associated with unresolved experiences that you may not be aware of. It's not necessary for you to remember the details of an experience to resolve it. You can be in a dreaming state and allow your body to move in resonance with the sound. There are three types of movement that may take place during a Human Tune In™: unwinding movements, communication movements, and sonic anchor movements. They are described below.

Unwinding Movements

Your body is intelligent; it holds somatic memories. Ideally, your body should naturally rise and fall with the sound waves. When something in you resists the waves, you may experience it as a pressure, as an uncomfortable feeling, or as an emotion or a thought. When this happens, let your body know that you are safe and then gradually let go of conscious control of your body. Letting go doesn't mean completely letting go. It means gradually letting go until you come to a place that is safe and your body is spontaneously moving. This is called unwinding.

To better understand unwinding movements, imagine that areas of pressure or constriction are part of your life that are "wound up like a rubber band." As you become aware of them and gradually relax, like a rubber band being freed up, you will naturally unwind. Unwinding movements should never be forced or planned. Just relax, trust your body and its intelligence, and allow it to naturally unwind. Normally, unwinding movements are unconscious micro-vibrational movements that you would not consciously notice. However, if you become aware of your body moving, you may notice a release, a slight pop, a twitch, or a deep breath. You may want to slightly shift your body, lift a shoulder, gently twist, rotate a leg, or you may want to stand up and make larger movements like walking, spinning, bending, or twisting. Be aware of the people around you and adjust your movements so as not to interfere with those around you.

Communication Movements

True spiritual voices are heard when the mind is in a state of deep absorption without conscious thought. While being in the sound, you may meet and have conversations with different beings. These beings are energy forms that appear in ways that are congruent with your reality structures. They can be parts of yourself, or of family members, people you know, or of people you don't know you are meeting in sonic space in the form of light beings, spirit guides, power animals, divine presences, or angels. When you communicate with others in normal conscious reality, your body and hands will naturally move to express something. The same sort of movement can happen during a sound healing concert. Although listeners may not speak out loud, it is OK for them to subvocalize. In other words, their lips and hands may move, but the sound is internal.

Sonic Anchor Movements

During a sound healing concert, you may want to bring your hand or hands to a specific place on your body. For example, if you want to work with your heart, place one hand or both on your heart. This is called a somatic anchor. Somatic anchors focus the sound waves moving through your body. After the concert, at any time, you can use your somatic anchor to recall the concert and to focus the sound vibrations on a specific area. (See Appendix C for how to create somatic anchors.)

TUNING YOUR BODY MOVEMENTS IN THE CONCERT SPACE

Be aware that every concert space has movement limitations. Be aware of and considerate of those around you when moving. It may be that you'll have to program yourself to make smaller movements or to bring your normal conscious mind back for a moment to find a safe place in the room to move. If it isn't possible to move around the room, then allow your body to make movements within your safe personal space that is defined by the room's limitations.

SOUND HEALERS

During a Human Tune In™ concert, the sound healers tap tuning forks and move around the room. Their intention is to create quality healing sound. They use their intuition to move their tuning forks in different patterns based on their sense of the energy in the room. At anytime they may tune into and direct sound to a group or an individual within the larger audience.

COMPOSITIONS

APOLLO'S LYRE uses a diatonic Pythagorean tuning system based on the natural division of a vibrating string as discovered by the Greek mathematician Pythagoras of Samos. Eight BioSonic tuning forks beginning at C256 cps, D, E, F, G, A, B, and ending at C512 cps are used to sound the archetype of Apollo's Lyre.

SEA SHELL HARMONICS uses eight BioSonic tuning forks tuned to the Fibonacci number series and designed to resonate with the pathway of consciousness from the sphenoid/pituitary axis to the pineal gland. Their purpose is to open new neural pathways and to empower creative visions. This aids in healing addictions and traumas. Intuitively, the Fibonacci number series can be seen as a spiraling sea shell. The center of the spiral is a still point through which we transition into alternate realities congruent with our intention. It is the archetype of Dorothy longing for "somewhere over the rainbow" and riding the vortex of a cyclone to the alternate reality of Oz where she discovers the courage to have a new creative vision with a heart.

MUSIC OF THE SPHERES This composition uses eleven tuning forks tuned to the planets. The ancient astronomers did not see the planets as fixed objects in space as we do today. They experienced them more like a lyre that was constantly creating music that affected everything on earth. They called it the law of cosmic sympatheia which was a direct expression of the Hermetic Axiom, "As above, then so below." For them, the planets were a cosmic muzak that always sounded in the background and affected all aspects of their lives.

The quest to hear the music of the planets continues to this day. When NASA's Voyager satellites picked up electromagnetic waves coming from the planets, they converted them to audible sounds and made them available on their web site. The sounds of the planets are subliminal. They can be made audible by converting their orbital cycles into frequencies and then raising those frequencies 29 to 38 octaves into the range of human hearing.

MERMAID DREAMS uses tuning forks tuned to the asteroids. The five main tunings are Chiron, Ceres, Pallas Athena, Juno, and Vesta. The asteroids were said to channel the mermaid voices of the muses. Tuning into the asteroids opens different channels of artistic creativity.

APPENDIX C

Anchoring and Values Visualization

Once you have determined the value and feeling tone of your visualization, it is then possible to establish an anchor. An anchor is a way of remembering and focusing the vibration of your visualization. There are three ways to use an anchor, somatically through the body, a spoken affirmation, and by humming. These can be worked with individually or can be combined to create an integrated anchor.

SOMATIC ANCHOR

The first type of anchor is a somatic anchor. To create a somatic anchor, tune into the feeling tone of your visualization and choose an area of your body to anchor the felt sense of the feeling tone. For example, you can touch your heart, hold your hands together, tap the top of your head, etc. To activate your somatic anchor, relax your rational mind, take a deep breath, and initiate your own unique somatic anchoring process. Once you have created your somatic anchor, you can use it any time. You can use it just before or when you are listening to sounds. You can use it before meditation. You can use it to direct a specific healing and/or to channel energy. For example, if you are concerned about a physical condition, i.e., your heart, place your hands over your heart and listen to the sound. Next, create a somatic anchor such as placing a hand on your heart. Athletes create and activate somatic anchors for themselves all the time. For example, golfers use somatic anchors to recall positive golf experiences. They may lift their thumb on and off the club just before they putt or touch their left shoulder just before they swing.

AFFIRMATION ANCHOR

The second type of anchor is a spoken or mentally repeated affirmation. **Affirmations are condensed statements of your values and are very effective methods of supporting positive change and wellness.** A research study conducted at Arizona State University suggested that the use of affirmations can decrease anxiety, lessen depression, and reduce addictive behaviors.[1] When properly constructed and used, affirmations are an important addition to the values visualization sound healing process. When constructing a values visualization affirmation based on the values you have identified, be sure to follow the suggested guidelines below.

a. Use words that are congruent with your values and the feeling tone or your visualization.

b. Modify or change words to resonate with your way of speaking.

c. Write affirmations in the present tense as though you are experiencing what you are affirming right now. It will not work to say, "In two weeks I will feel better." It will work to say, "I am healing" or "I feel better right now."

d. Focus on the positive and avoid using negative terms, i.e., if you are having financial challenges, it will not work to say, "I am not going to spend more money." Instead, visualize your future and see and feel yourself as having just the right amount of money to express your values and to create a positive affirmation like, "I am wealthy and prosperous."

e. Repeat your affirmation out loud whenever you can. Allow the sound of your voice to become a sound healing instrument. Advertisements on television and radio always repeat their ads. The same holds true for affirmations. It is OK to repeat an affirmation in many different ways with different sounds and voices. Two of the most recommended times to repeat your sound affirmation process is at night before you go to sleep and in the morning when you wake up.

f. If you use a pre-written affirmation, it is important to evaluate it by going through the above five steps. It may work to write it down, or you may choose to change some of the wording. The affirmation could also inspire you to write a new affirmation, one that is a more finely tuned expression of your values.

Notes About Affirmations

There are thousands of affirmations available in books and on the internet. When you research affirmations, it is important to look at them as though you were writing them. When an affirmation gets your attention, evaluate it by going through the affirmation guidelines. You can write it down or feel free to change any words. Here are some more affirmation ideas. Feel free to change them, share them, and use them to create new affirmations.

- I put this into the light for my highest values.
- I put this into the light and allow the sound to guide me to the best possible place for healing.

- I put this into the light to illuminate creative solutions for my highest values.
- I am a vibrating being of light listening to dream secrets of perfect thoughts.
- I am a tuning fork vibrating illuminated healing waves through my body.
- I am a dolphin swimming in dreamtime to the lost city of Atlantis
- I am learning from the best experiences of my life to resolve today's challenges.
- I will express myself in exactly the right way every day and in every way.
- I am a creative being with creative solutions.
- Challenges are my new opportunities.
- I ready to face the unknown with strong positive thoughts.
- I am in a deep magical well revealing visions of my true self.
- I am listening to myself. I am always discovering.
- I am valuable. I am kind and patient with myself.
- I am learning to express myself in new and creative ways.
- I am a vibrant creative human being.

HUMMING ANCHOR

The third type of anchor is humming. Everyone hums. We hum our favorite tunes. We spontaneously hum if we have a question and/or are solving a problem. We hum when we are happy. Sometimes we "just hum." When using humming as an anchor:

a. Tune into the feeling tone of your visualization.
b. Hum a sound that resonates with your feeling tone.

or

a. Tune into the feeling tone of your visualization.
b. Listen to different music until you discover a tune or song that resonates with your feeling tone.
c. Tune into your feeling tone and hum the music.

How to Use Anchors with Tuning Forks

a. Be in a safe place.

b. Look at your written affirmation, and tune into and "feel" your words beyond their intellectual meaning.

c. Speak and/or think your affirmation with a confident voice.

d. Tap C & G tuning forks, bring them to your ears, enter a deep listening state, and go into the sound.

e. When the sound goes into silence, take a moment to activate your somatic anchor and then allow your normal conscious mind to surface.

APPENDIX D
Five Element Hand Shaking Exercise

When we shake hands, we are making contact with living, pulsating, rhythmic beings. Each being has his own unique self. If we are quiet inside and listen, we can sense their essence through the touch of the handshake. This unique essence we sense is their consciousness expressing itself through the elements.

EARTH SHAKE

Grasp another person's hand slowly, and grip it securely and solidly, as though you were hugging the hand. Your pressure should not be too tight or too loose. There is not much movement. Just an "earthy still security" that says, "I am here. I have you. All is safe and secure. You can count on me."

WATER SHAKE

Grasp the other hand in a "feely touchy way." The movements should be wave-like and the grip can range from slightly tight to very loose. The grip is always changing. More importantly, the hand shake is not limited to the hands. Touch the other person with your non-shaking hand, perhaps on the shoulder, or on the wrist while shaking. You are saying to the person I am bonding with you. We are together and attached. We are flowing together.

FIRE SHAKE

Directly look the person in the eye. Reach out in a direct way. Squeeze their hand securely and move up and down in a fast focused motion. You are saying to the person I am here and let's get on with it—no time to mess around—this is all about action. Say, "Good to meet you and see you later." No time for talking. I am on to the next person.

AIR SHAKE

Grasp the other person's hand quickly and with a very light touch. Shake and then slightly let go and shake again and slightly let go and shake again. Do not directly shake. Each shake should be a little different and even erratic. It is as if you are saying to the person, "I am here and, at the same time, I am at many other places and thinking about other things." Or, "There is so much and let's not get bound up in anything specific. But I am here, at least for the moment."

ETHER SHAKE

Be very still inside. Acknowledge the other person with your presence. Hold your hands together and bow to the other person. Create just the right amount of space and subtly acknowledge the other person with an eye gesture or a minimal body movement.

APPENDIX E
Element Advertisement Examples

Fire (with only sound of breath): Nike Breath Commercial
   https://www.youtube.com/watch?v=kdjchJw16mI

Earth with Water: Lost Dog Budweiser Commercial
   https://www.youtube.com/watch?v=TPKgC8KPBMg

Earth to Fire: Viagra Commercial
   https://www.youtube.com/watch?time_continue=4&v=UYvC1aEKBEM

Earth and Fire: Reebok Freak Show Commercial
   https://www.youtube.com/watch?v=Ijhk5Emd6Ig

Water to Fire: Best VW Commercial Ever
   https://youtu.be/WxjP4tOuI6E

Earth: Lunesta Commercial
   https://www.youtube.com/watch?v=kWdt7yn2BoU

Earth and Water ending in Fire: Leo Beer Commercial
   https://www.youtube.com/watch?v=Ojdq_TRlY4o

Air and Fire: Volkswagen Beetle 2011 Super Bowl Commercial
   https://www.youtube.com/watch?v=-NGN4J6F_vI

APPENDIX F

Hypnosis and Element Integration Examples

Over the years my patients have inspired me to create hypnosis recordings that were inspired by their healing goals. The background music and sounds are based on different elemental mixes that resonate with different healing and desired outcomes. They are available on my web site, www.biosonics.com, as free downloads for anyone to use and learn from. The list of the recordings is given according to their topic and the element carrier music that is used in the background.

- Motivation (*Fire and Earth*)
- Sweet Dreams (*Earth and Water*)
- Relationships (*Water*)
- Butterfly (*Water with Air*)
- Creating (*Air and Water*)
- Body Healing (*Earth with Water*)
- Finding Your Life Partner (*Water and Earth*)
- Relationships (*Earth with Water and some Air*)
- Success (*Air*)
- Challenges (*Earth*)
- Dream Dolphin Trance #1 (*Earth with Air*)
- Dream Dolphin Trance #2 (*Earth with Water*)

Endnotes

FOREWORD

[1] A. Chiesa and A. Serretti, "A Systematic Review of Neurobiological and Clinical Features of Mindfulness Meditations," *Psychological Medicine* 40.08 (2009): 1239-252. Web.

[2] Gerhard Roth (00021) and Ursula Dicke (00021), "Evolution of Nervous Systems and Brains," Springer, Springer Berlin Heidelberg, 01 Jan. 1970. Web. 27 May 2017. http://link.springer.com/chapter/10.1007/978-3-642-10769-6_2.

[3] Ibid.

[4] James E. Peck. "The Development of Hearing, Part I," *American Academy of Audiology*, n.d. Web.

[5] James E. Peck. "Evolution of Nervous System," Academic Press, 16 Dec. 2016, 2nd Edition. Web. 27 May 2017. https://www.elsevier.com/books/evolution-of-nervous-systems/kaas/978-0-12-804042-3.

INTRODUCTION

[1] E. L. Deci and R. M. Ryan, "The 'what' and 'why' of goal pursuits: Human needs and the self-determination of behavior," *Psychological Inquiry* 11 (2000): 227–268.

[2] B. S. Wiese, "Successful pursuit of personal goals and subjective well-being," in *Personal Project Pursuit: Goals, Action and Human Flourishing*, eds. B. R. Little, K. Salmela-Aro, and S. D. Phillips (Hillsdale, NJ: Lawrence Erlbaum, 2007): 301–328.

[3] John Beaulieu and David Perez-Martinez, "Sound Healing Theory and Practice," in *Nutrition and Integrative Medicine: A Primer for Clinicians*, ed. Aruna Bakhru (Boca Raton, FL: CRC Press, 2018).

[4] John Beaulieu, *Human Tuning* (Stone Ridge, NY: BioSonic Enterprises, Ltd., 2010), 71.

[5] John Beaulieu, 2002. "Otto 128 Tuning Fork and Nitric Oxide Response," unpublished raw data set.

[6] John Beaulieu, "The Perfect Fifth: The Science and Alchemy of Sound," *Rose+Croix Journal* (2017), http://www.rosecroixjournal.org.

SECTION ONE
PART 1: Sacred Sound Science

1. John Beaulieu and David Perez-Martinez, "Sound Healing Theory and Practice," in *Nutrition and Integrative Medicine: A Primer for Clinicians*, ed. Aruna Bakhru (Boca Raton, FL: CRC Press, 2018).

2. Werner Hesinberg, *Physics and Philosophy* (New York, NY: Harper & Row, Inc., 1962).

3. Argone National Laboratory, "Acoustic Levitation" (2012), https://www.youtube.com/watch?v=669AcEBpdsY.

4. Weiyu Ran, Steven Fredericks, and John R. Saylor, "Shape oscillation of a levitated drop in an acoustic field," (Cornell University Library, 2013), https://arxiv.org/find/physics/1/au:+Saylor_J/0/1/0/all/0/1.

5. Christian B. Anfinsen, "Principles that Govern the Folding of Protein Chains," *Science* 181 (July 20, 1973).

6. Katherine Kornei, "Listen to the music of protein folding," *Science* (Nov. 18, 2016), http://www.sciencemag.org/news/2016/11/listen-music-proteins-folding.

7. Karen Barad, *Halfway: Quantum Physics and the Entanglement of Meaning* (Duke University Press, 2007), 254.

8. Terry Riley, "The Trinity of Eternal Music," in *Sound and Light: La Monte Young & Marian Zazeela*, eds. William Duckworth and Richard Fleming (Cranbury, NJ: Associate University Presses, 1996).

9. Juan Miguel Marin, "'Mysticism' in quantum mechanics: the forgotten controversy," *European Journal of Physics* 30 (2009): 807–822.

SECTION ONE
PART 2: Mindful Listening

1. Jeffrey B. Rosen and Melanie P. Donley, "Animal studies of amygdala function in fear and uncertainty: Relevance to human research," *Biological Psychology* 73 (July 2006): 49–60.

2. John Beaulieu, *Music and Sound in the Healing Arts* (Barrytown NY: Station Hill Press, 1987), 120–123.

3. Horatiu Boeriu, "How the Perfect Car Door Sound Is Made for BMW," *BMW Blog* (Dec. 2014), http://www.bmwblog.com/2014/12/22/perfect-car-door-sound-made-bmw/.

4. "Audi Creates Sound for Electric Car," *Copperworld* (2012), https://www.youtube.com/watch?v=nY2wB_PCEm8.

5. S. D. Edwards, "Influence of HeartMath Quick Coherence Technique on Psychophysiological Coherence and Feeling States," *African Journal for Physical Activity and Health Sciences* 22, (Dec. 2016): 1006–1018.

6. Richard P. Brown and Patricia L. Gerbarg, "Sudarshan Kriya Yogic Breathing in the Treatment of Stress, Anxiety, and Depression: Part I Neurophysiologic Model," *The Journal of Alternative and Complementary Medicine* II, no. 1 (2005).

7. John Beaulieu, *Music and Sound in the Healing Arts* (Barrytown, NY: Station Hill Press, 1987), 17.

SECTION ONE
PART 3: Be Like a Child

[1] John Cage, *Silence* (Hanover, NH: University Press of New England, 1961), 4.

[2] Henry Cowell used string piano sounds in his compositions *Aeolian Harp* and *The Banshee*.

[3] John Cage's best known compositions for prepared piano are *Sonata* and *Interludes*.

[4] Iannis Xenakis, *Formalized Music: Thought and Mathematics in Composition*, 2nd ed. (Pendragon Press, Jan. 1992).

[5] John Beaulieu, *Calendula: Sound Healing with Tuning Forks*, CD (Stone Ridge, NY: BioSonic Enterprises, Ltd.).

[6] John Beaulieu, *Apollo's Lyre: A Modern Adventure in Ancient Cosmic Harmony*, CD (Stone Ridge, NY: BioSonic Enterprises, Ltd., 2016).

SECTION ONE
PART 4: Listen Like a Scientist

[1] Wolfgang Pauli, *Writings on Physics and Philosophy*, eds. Charles P. Enz and Karl von Meyenn (New York: Springer-Verlag New York, 1994), 259.

[2] "Phenomenology," *Stanford Encyclopedia of Philosophy* (Nov. 2003), www.plato.stanford.edu/entries/phenomenology.

[3] Fritz Perls, *Gestalt Therapy Verbatum* (Gouldsboro, ME: The Gestalt Journal Press, 1992).

[4] John Beaulieu, *Human Tuning* (Stone Ridge, NY: BioSonic Enterprises, Ltd., 2010), 4–7.

[5] James Hablin, "Bone Fone, the Terror!" *The Atlantic*, https://www.theatlantic.com/health/archive/2013/09/bone-fone-the-terror/279474/.

[6] Daniel K. Stat, "Double-chambered whistling bottles: A unique Peruvian pattern form" (1974), http://www.peruvianwhistles.com/journ-transpersonal.html.

[7] John Beaulieu, *Calendula: A Suite for Pythagorean Tuning Forks*, CD (Stone Ridge, NY: BioSonic Enterprises, Ltd., 1998).

[8] John Beaulieu, *Spirit Whistles: A Suite for Overtone Whistles*, CD (Stone Ridge, NY: BioSonic Enterprises, Ltd., 2000).

[9] Firas Khatib, Seth Cooper, Michael D. Tyka, Kefan Xu, Illya Makedon, Zoran Popovic, Dave Baker, and Foldit Players, "Algorithm discovery by protein folding game players," *Proceedings of National Academy of Sciences* 108, no. 47 (June 29, 2011), http://www.pnas.org/content/108/47/18949.full.

SECTION TWO
PART 1: Visualization

[1] Christopher C. Berger and Henrik H. Ehrsson, "Mental Imagery Changes Multisensory Perceptions," *Current Biology* (June 2013), 10.1016/j.cub.2013.06.012, http://www.cell.com/current-biology/abstract/S0960-9822(13)00703-3.

[2] Pelle Cass, "Wonderland of Dough: Photos of American Convenience Stores that Have Sold Million-Dollar Lottery Tickets," *Feature Shoot* (2014), http://www.featureshoot.com/2014/07/wonderland-of-dough-photos-of-american-convenience-stores-that-have-sold-million-dollar-lottery-tickets/.

[3] Jia Qi, Shiliang Zhang, Hui-Ling Wang, Huikun Wang, Jose de Jesus Aceves Buendia, Alexander F. Hoffman, Carl R. Lupica, Rebecca P. Seal, and Marisela Morales, "A glutamatergic reward input from the dorsal raphe to ventral tegmental area dopamine neurons," *Nature Communications* (2014): 5: 5390 DOI: 10.1038/ncomms6390.

[4] National Institute of Drug Abuse, "Understanding Drug Abuse and Addiction: What Science Says," https://www.drugabuse.gov/publications/teaching-packets/understanding-drug-abuse-addiction/section-i/4-reward-pathway.

[5] Scott Hankins, Mark Hoekstra, and Paige M. Skiba, *The Review of Economics and Statistics* (MIT Press Journals, 2011), http://www.mitpressjournals.org/doi/abs/10.1162/REST_a_00114.

SECTION TWO
PART 2: Values

[1] Benedicte Apouey and Andrew E. Clark. (2015) "Winning Big but Feeling No Better? The Effect of Lottery Prizes on Physical and Mental Health," *Health Economics* 24:5516–538. Online publication date: Feb. 18, 2014. No index cross ref.

[2] Howard Kirschenbaum, *Values Clarification in Counseling and Psychotherapy* (New York, NY: Oxford University Press, 2013).

SECTION TWO
PART 3: Feeling Tones

[1] William Gray, "Understanding creative thought processes: An early formulation of the emotional-cognitive structure theory," *Man-Environment Systems* 9 (1979): 314.

[2] Manfred Clynes, *Sentics: The Touch of the Emotions* (New York, NY: Australia Prism Press Ltd., 1989).

[3] Hans Jenny, *Cymatics: The Structure and Dynamics of Waves and Vibrations* (Basel, Switzerland: Basilius Press, 1974).

[4] D. D. Palmer, *Text-Book of the Science, Art and Philosophy of Chiropractic for Students and Practitioners* (Portland, OR: Portland Printing House Company, 1910:7).

[5] Selen Atasoy, Isaac Donnelly, and Joel Pearson, "Human brain networks function in connectome-specific harmonic waves," *Nature Communications* (Jan. 2016), http://www.nature.com/articles/ncomms10340#content.

[6] R. T. Canolly, E. Edwards, S. S. Dalal, M. Soltani, S. S. Nagarajan, H. E. Kirsch, N. M. Barbaro, and R. T. Knight, "High gamma power is phase-locked to theta oscillations in human neocortex," *Science* (2006): 1626-8, 15:31.

[7] György Buzsàki, "Theta Oscillations in the Hippocampus," *Neuron* 33 (Jan 2002): 325340.

[8] Peter D. Bruza, Jerome R. Busemeyer, and Zheng Wang, "Quantum cognition: A new theoretical approach to psychology," *Cell Symposia* 19 (July 2016): 383–393.

[9] Robert P. Bywater and Jonathan N. Middleton, "Melody discrimination and protein fold classification," *Bioinformatics* 2 (Oct. 2018).

[10] Mark Reybrouck and Elvira Brattico, "Neuroplasticity Beyond Sounds: Neural Adaptations Following Long-Term Musical Aesthetic Experiences," *Brain Sciences* (2015): 69–91.

[11] Michael F. Barnsley, Robert L. Devaney, Benoit B. Mandelbrot, Heinz-Otto Peitgen, Dietmar Saupe, and Richard F. Voss, *The Science of Fractal Images* (Springer-Verlag, 1988).

[12] Gary Ferraro and Susan Andreatta, *Cultural Anthropology: An Applied Perspective* (Wadsworth, CA, 2012), 355.

[13] John Neihardt, *Black Elk Speaks* (Albany NY: State University of New York Press, 2008).

SECTION TWO
PART 4: Sound and Values Visualization

[1] Michael W. Reimann, Max Nolte, Martina Scolamiero, Katharine Turner, Rodrigo Perin, Giuseppe Chindemi, Pawel Dlotko, Ran Levi, Kathryn Hess, and Henry Markram, "Cliques of Neurons Bound into Cavities Provide a Missing Link Between Structure and Function," *Frontiers in Computational Neuroscience* (June 12, 2017).

[2] Stewart Hamereoff and Roger Penrose, "Orchestrated reduction of quantum coherence in brain microtubules: A model for consciousness," *Mathematics and Computers in Simulation* 40 (1996): 453–480.

[3] Stewart Hammeroff, "Is Your Brain Really a Computer, or Is It a Quantum Orchestra?," *Scientific Association for the Study of Time in Physics and Cosmology Speaker Series*, https://timeincosmology.com/2015/10/12/hameroff/.

[4] Cathal O'Connell, "Topology explained—and why you're a donut," *Cosmos* (Oct. 2016), https://cosmosmagazine.com/physics/what-is-topology.

[5] Dane Rudyhar, *The Magic of Tone and the Art of Music* (Boulder, CO: Shambhala, 1982).

[6] "Alive Inside, Man in Nursing Home Reacts to Hearing Music from His Era," *Music and Memory* (Nov. 2011), https://www.youtube.com/watch?v=fyZQf0p73QM&t=172s.

[7] Thea K. Beaulieu, *Moving with the Elements* (Stone Ridge NY: BioSonic Enterprises, Ltd., 2016).

[8] Jon D. Morris and Mary Anne Boone, "The Effects of Music on Emotional Response, Brand Attitude, and Purchase Intent in an Emotional Advertising Condition," *NA—Advances in Consumer Research* 25, eds. Joseph W. Alba and J. Wesley Hutchinson (Provo, UT: Association for Consumer Research, 1988): 518–526.

SECTION THREE
Sound Healing and Values Visualization Practice

[1] Lee N. Robins, *The Vietnam Drug Use Returns* (St. Louis MO: Washington School of Medicine, 1973), http://prhome.defense.gov/Portals/52/Documents/RFM/Readiness/DDRP/docs/35%20Final%20Report.%20The%20Vietnam%20drug%20user%20returns.pdf.

[2] Julia K. Boehm, Laura D. Kubzansky, and Suzanne C. Segerstrom, "Positive Psychological Functioning and the Biology of Health," *Social Personality Psychology Compass* 9 (Dec. 2015): 645–660.

SECTION FOUR
Sound Musings

[1] Manly P. Hall, *The Therapeutic Value of Music* (Los Angeles, CA: Philosophical Research Society, 1982).

[2] Ibid., 1–3.

[3] Michael S. Schneider, "The Mathematical Archetypes of Nature, Art, and Science," *A Beginner's Guide to Constructing the Universe* (New York: Harper, 1994): 237–240.

[4] John Beaulieu, *Music and Sound in the Healing Arts* (Barrytown, NY: Station Hill Press, 1987), 45.

[5] Ibid., 46.

[6] Manly P. Hall, "The Pythagorean Theory of Music and Color," in *The Secret Teachings of All Ages* (CA: Philosophical Research Society, 1988), lxxxi–lxxxv.

[7] William Smith and Samuel Cheetham, *A Dictionary of Christian Antiquities* (London: John Murray, 1875), 550.

[8] John Beaulieu, *Human Tuning* (Stone Ridge, NY: BioSonic Enterprises, 2010), 40–52.

[9] Swami Rama, "Darshan Notes" (lecture presented at the Himalayan Institute, Honesdale, PA, 1991).

[10] Lao Tzu, *Tao Te Ching* (CA: Sacred Books of the East, Vol. 39, 1891), 42.

[11] T. S. Eliot, "Burnt Norton," *The Four Quartets* (Orlando, FL: Harcourt, 1973).

[12] Warren D. Anderson, *Music and Musicians in Ancient Greece* (Ithaca, NY: Cornell University Press, 1997).

[13] Manly P. Hall, *Melchizedek and the Mystery of Fire* (Los Angeles, CA: Philosophical Research Society, 1996).

[14] Hans Erhard Laurer, *Cosmic Music: The Evolution of Music Through Changes in Tone Systems* (VT: Inner Traditions, 1989), 197–200.

[15] John Beaulieu, *Human Tuning* (Stone Ridge, NY: BioSonic Enterprises, 2010), 47.

[16] Alexander Roob, *Alchemy & Mysticism* (New York: Taschen, 1997), 126–145.

[17] Manly P. Hall, *The Manly P. Hall collection of alchemical manuscripts* (CA: Getty Research Institute), http://www.getty.edu/research/.

[18] Nicholas Goodrick-Clarke, ed., *Paracelsus: Essential Readings* (Berkeley, CA: North Atlantic Books, 1999), 75.

[19] Silvia Nakkach, *Free Your Voice* (CO: Sounds True, 2012), 133–165.

[20] Silvia Nakkach, "Song the Healers Hear," *Medicine Melodies*, CD (CO: Sounds True, 2011).

[21] John Beaulieu, *Human Tuning* (Stone Ridge, NY: BioSonic Enterprises, 2010).

[22] John Beaulieu, *Apollo's Lyre: A Modern Adventure in Ancient Cosmic Harmony.* (NY: BioSonic Enterprises, Ltd., 2016).

[23] Rudolf Haase, "Harmonics and Sacred Traditions," *Cosmic Music*, ed. Joscelyn Godwin (VT: Inner Traditions, 1989), 91–131.

[24] John Beaulieu and David Perez-Martinez, "Sound Healing Theory and Practice," in *Nutrition and Integrative Medicine: A Primer for Clinicians,* ed. Aruna Bakhru (Boca Raton, FL: CRC Press, 2018).

[25] Ibid.

[26] John Beaulieu, *Human Tuning* (Stone Ridge, NY: BioSonic Enterprises, 2010), 40–52.

[27] John Beaulieu and David Perez-Martinez, "Sound Healing Theory and Practice," in *Nutrition and Integrative Medicine: A Primer for Clinicians,* ed. Aruna Bakhru (Boca Raton, FL: CRC Press, 2018).

[28] Ingmar Leshmann and Alfred S. Posamentier, *The (Fabulous) Fibonacci Numbers* (NY: Prometheus Books, 2007).

[29] Carolyn Boroden, *Fibonacci Trading* (NY: McGraw Hill, 2008).

[30] The original drawings used in this case study were gifted to the author, and the name of the patient has been changed.

[31] John Beaulieu, *Music and Sound in the Healing Arts* (Barrytown, NY: Station Hill Press, 1987), 53.

[32] John Beaulieu, *Human Tuning* (Stone Ridge, NY: BioSonic Enterprises, 2010), 40–52.

[33] Xu Weiming, Li Zhi Liu, Marilena Loizidou, Mohamed Ahmed, and Ian Charles, "The role of nitric oxide in cancer," *Cell Research* 12 (2002).

[34] Mona Lisa Chanda and Daniel J. Levitin, "The Neurochemistry of Music," *Trends in Cognitive Sciences* 17.4 (2013), 179–193. Web.

[35] John Beaulieu, George B. Stefano, Elliott Salamon, and Minsun Kim, "Sound Therapy Induced Relaxation: Down Regulating Stress Processes and Pathologies," *Medical Science Monitor* (2003).

[36] Daniel J. Levitin and Mona Lisa Chanda, "The Neurochemistry of Music," *Trends in Cognitive Sciences* 17.4 (2013): 179–193. Web.

[37] C. L. Pelletier, "The Effect of Music on Decreasing Arousal Due to Stress: A Meta-Analysis," *Journal of Music Therapy* 41.3 (2004): 192–214.

[38] P. Kreutz, "Effects of choir singing and listening on secretory immunoglobulin A, cortisol, and emotional state," *Journal of Behavioral Medicine* 27 (2006): 171–179.

[39] Herbert Benson, Gregory L. Fricchione, Brian T. Slingsby, and George B. Stefano, "The placebo effect and relaxation response: neural processes and their coupling to constitutive nitric oxide," *Brain Research Reviews* (2001): 1–19.

[40] Hans Selye, *The Stress of Life* (NY: McGraw-Hill, 1978), 43.40.

APPENDIX C
Anchoring and Values Visualization

[1] Richard T. Kinner, Christy Hofsess, Rick Pongratz, and Christina Lamberet, *Psychology and Psychotherapy: Theory, Research, and Practice* 82, issue 2 (June 2009): 153–169, DOI: 10.1248/14760830X3804 18.

Bibliography

"Acoustic Levitation." Argone National Laboratory. https://www.youtube.com/watch?v=669AcEBpdsY. Sept. 2012.

"Alive Inside, Man in Nursing Home Reacts to Hearing Music from His Era." *Music and Memory*. https://www.youtube.com/watch?v=fyZQf0p73QM&t=172s. Nov. 2011.

Anderson, Warren D. *Music and Musicians in Ancient Greece*. Ithaca, NY: Cornell University Press, 1997.

Bandyopadhyay, Anirban, Satyajit Sahu, Subrata Ghosh, and Daisuke Fujita. "Live visualizations of single isolated tubulin protein self-assembly via tunneling current: effect of electromagnetic pumping during spontaneous growth of microtubule." *Scientific Reports*. 2014.

Barad, Karen Michelle. *Meeting the Universe Halfway*. Durham, NC: Duke University Press, 2007.

Barad, Karen. *Halfway: Quantum Physics and the Entanglement of Meaning*. Durham, NC: Duke University Press, 2007.

Beaulieu, John, George B. Stefano, Elliott Salamon, and Minsun Kimm. "Sound Therapy Induced Relaxation: Down Regulating Stress Processes and Pathologies." *Medical Science Monitor*. 2003.

Beaulieu, John. *Music and Sound in the Healing Arts*. Barrytown, NY: Station Hill Press, 1987.

Beaulieu, John. *Human Tuning: Sound Healing with Tuning Forks*. Stone Ridge, NY: BioSonic Enterprises, Ltd., 2010.

Beaulieu, John, and David Perez-Martinez. "Sound Healing Theory and Practice." *Nutrition and Integrative Medicine: A Primer for Clinicians*. Editor: Aruna Bakhru. Boca Raton, FL: CRC Press, 2018.

Beaulieu, John. "The nature of healing sound and its design." *Psychology's New Design Science and The Reflective Practitioner*. Editors: Susan Imholz and Judy Sachter. River Bend, NC: LibraLab Press, 2018.

Beaulieu, John. "The Perfect Fifth: The Science and Alchemy of Sound." *Rose+Croix Journal*. http://www.rosecroixjournal.org. 2017.

Beaulieu, John. *Polarity Therapy Workbook,* 2nd Ed. Stone Ridge, NY: BioSonic Enterprises, 2015.

Beck, Friedrich. "Synaptic Quantum Tunneling in Brain Activity." *NeuroQuantology*, Vol. 6, No. 2. 2008.

Becker, Rollin E. *Life in Motion*. Portland, OR: Rudra Press, 1997.

Benson, H., and W. Proctor. *The Break-Out Principle*. New York, NY: Scribner, 2003.

Berger, Christopher C., and Henrik H. Ehrsson. "Mental Imagery Changes Multisensory Perceptions." *Current Biology*. 10.1016/j.cub.2013.06.012. http://www.cell.com/current-biology/abstract/S0960-9822(13)00703-3. June 2013.

Bertalanffy, Ludwig von. *General System Theory: Foundations, Development, Applications*. New York, NY: George Braziller, 1968.

Bhoria, Renu, and Panipat Swati Gupta. "A Study of the Effect of Sound on EEG." *International Journal of Electronics and Computer Science Engineering*. 1956.

Boeriu, Horatiu. "How the Perfect Car Door Sound Is Made for BMW." *BMW Blog*. http://www.bmwblog.com/2014/12/22/perfect-car-door-sound-made-bmw/. Dec. 2014.

Bohm, D., and B. Hiley. "On the Intuitive Understanding of Nonlocality as Implied by Quantum Theory." *Foundations of Physics*, Vol. 5. 1975.

Brunet, P., and J. H. Snoeijer. "Star-drops formed by periodic excitation and on an air cushion." *European Physical Journal Special Topics*. Springer-Verlag, 2011.

Bywater, P., Jonathan N. Middleton. "Melody discrimination and protein fold classification Heliyon." *Bioinformatics*. http://www.heliyon.com/article/e00175. Oct. 2016.

Cage, John. *Silence*. Middletown, NY: Wesleyan University Press, 1965.

Cass, Pelle. "Wonderland of Dough: Photos of American Convenience Stores that Have Sold Million-Dollar Lottery Tickets." *Feature Shoot*. http://www.featureshoot.com/2014/07/wonderland-of-dough-photos-of-american-conveniencestores-that-have-sold-million-dollar-lottery-tickets. 2014.

Castaneda, Carlos. *The Teachings of Don Juan: A Yaqui Way of Knowledge*. New York, NY: Simon & Schuster, 1968.

Cooke, J. P. *The Cardiovascular Cure*. New York, NY: Broadway Books, 2002.

Cousto, Hans. *The Cosmic Octave: Origin of Harmony*. Mendocino, CA: Life Rhythm, 1988.

Eliot, T. S. "Burnt Norton." *The Four Quartets*. Orlando, FL: Harcourt, 1973.

Deci, E.L., and Ryan, R. M., "The 'what' and 'why' of goal pursuits: Human needs and the self-determination of behavior." *Psychological Inquiry* 11. 2000.

Doczi, Gyorgy. *The Power of Limits: Proportional Harmonies in Nature, Art, & Architecture*. Boston, MA: New Science Library, 1982.

Fountain, H. "Discovering the Tricks of Fireflies: Summertime Magic." *New York Times*. July 3, 2001.

Fredericks, S., W. Ran, and J. R. Saylor. *Shape Oscillation of a Levitated Drop in an Acoustic Field*. Clemson University Department of Mechanical Engineering Clemson. https://www.youtube.com/watch?v=4z4QdiqP-q8. 2013.

Garland, Trudi, and C. Kahn. *Math and Music: Harmonious Connections*. Parsippany, NJ: Dale Seymour Publications, 1995.

Goldman, Jonathan, and Goldman Andi. *The Humming Effect*. Rochester, VT: Healing Arts Press, 2017.

Goodchild van Hilten, Lucy. "Listening to proteins by turning data into music." *Elsevier*. https://www.elsevier.com/connect/listening-to-proteins-by-turning-data-into-music. 2016.

Goodrick-Clarke, Nicholas, ed. *Paracelsus: Essential Readings*. Berkeley, CA: North Atlantic Books, 1999.

Haase, Rudolf. "Harmonics and Sacred Traditions." *Cosmic Music*. Editor: Joscelyn Godwin. Rochester, VT: Inner Traditions, 1989.

Hablin, James. "Bone Fone, the Terror!" *The Atlantic*. https://www.theatlantic.com/health/archive/2013/09/bone-fone-the-terror/279474/.

Hall, Manly P. *The Therapeutic Value of Music*. Los Angeles, CA: The Philosophical Research Society, 1955.

Hall, Manly P. "The Pythagorean Theory of Music and Color." *The Secret Teachings of All Ages*. CA: Philosophical Research Society, 1988.

Hall, Manly P. *Melchizedek and the Mystery of Fire*. Los Angeles, CA: Philosophical Research Society, 1996.

Hall, Manly P. *The Manly P. Hall Collection of Alchemical Manuscripts*. CA: Getty Research Institute. http://www.getty.edu/research.

Hamereoff, Stewart, and Roger Penrose. "Orchestrated reduction of quantum coherence in brain microtubules: A model for consciousness." *Mathematics and Computers in Simulation*. 1996.

Hammeroff, Stewart. "Is Your Brain Really a Computer, or Is It a Quantum Orchestra?" *Scientific Association for the Study of Time in Physics and Cosmology*. Speaker Series. https://timeincosmology.com/2015/10/12/hameroff.

Jenny, Hans. *Cymatics: The Structure and Dynamics of Wave Phenomena and Vibrations*. Two-volume compilation re-issued 2001. Eliot, ME: MACROmedia Publishing, www.cymaticsource.com, 1974.

Qi, Jia, Shiliang Zhang, Hui-Ling Wang, Huikun Wang, Jose de Jesus Aceves Buendia, Alexander F. Hoffman, Carl R. Lupica, Rebecca P. Seal, and Marisela Morales. "A glutamatergic reward input from the dorsal raphe to ventral tegmental area dopamine neurons." *Nature Communications*. 2014.

Kays, Jill L., Robin A. Hurley, and Katherine H. Taber. "The dynamic brain: neuroplasticity and mental health." *The Journal of Neuropsychiatry and Clinical Neurosciences*. Web. 2012.

Khatib, Firas, Seth Cooper, Michael D. Tyka, Kefan Xu, Illya Makedon, Zoran Popovic, Dave Baker, and Foldit Players. "Algorithm discovery by protein folding game players." *Proceedings of National Academy of Sciences*. http://www.pnas.org/content/108/47/18949.full. June 2011.

Khan, Hazrat Inayat. *Music*. New York, NY: Samuel Weiser, 1962.

Klossowski de Rola, Stanislas. *Alchemy: The Secret Art*. London: Thames & Hudson, 1973.

Khanna, Madhu. *Yantra: The Tantric Symbol of Cosmic Unity*. London: Thames & Hudson, 1997.

Kornel, Katherine. "Science Listen to the music of proteins folding." *Science*. http://www.sciencemag.org/news/2016/11/listen-music-proteins-folding. Nov. 2016

Kotchoubey, Boris, Yuri G. Pavlov, and Boris Kleber. *Music in Research and Rehabilitation of Disorders of Consciousness: Psychological and Neurophysiological Foundations*. Institute for Medical Psychology and Behavioural Neurobiology, University of Tübingen, Tübingen, Germany.

Lauer, Hans Erhard. *Cosmic Music: The Evolution of Music Through Changes in Tone Systems*. Rochester, VT: Inner Traditions, 1989.

Lauterwasser, A. *Water Sound Images: The Creative Music of the Universe*. Newmarket, NH: Macromedia Publishing, 2002.

LaViolette, Paul A. "Thoughts about thoughts about thoughts: The emotional-perceptive cycle theory." *Man-Environment Systems 9*. 1979.

Levitin, Daniel J., and Mona Lisa Chanda. "The Neurochemistry of Music." *Trends in Cognitive Sciences* 17.4. 179-93. https://daniellevitin.com/levitinlab/articles/2013-TICS_1180.pdf. 2013.

Loewy, Joanne. "Integrating music, language and the voice in music therapy." *Voices: A World Forum for Music Therapy*. 2004.

Magoun, Harold I. *Osteopathy in the Cranial Field*. Kirksville, MO: Journal Printing, 1976.

Maratos, A., M. J. Crawford, and S. Procter. "Music Therapy for Depression: It Seems to Work, but How?" *The British Journal of Psychiatry*. http://bjp.rcpsych.org/content/199/2/92

Marin, Juan Miguel. "'Mysticism' in quantum mechanics: the forgotten controversy." *European Journal of Physics*. 2009.

Maslow, Abraham. *The Further Reaches of Human Nature*. New York, NY: Penguin Books, 1993.

Maugham, Somerset. *The Razor's Edge*. New York, NY: Vintage Books, 2003.

Morris, Jon D., and Mary Anne Boone. "The Effects of Music on Emotional Response, Brand Attitude, and Purchase Intent in an Emotional Advertising Condition." *NA—Advances in Consumer Research 25*.

Hutchinson, Wesley. Provo, UT: Association for Consumer Research, 1998.

Münte, Thomas F., Eckart Altenmüller, and Jäncke Lutz. 2002. *The Musician's Brain as a Model of Neuroplasticity*. Nature Publishing Group, Vol. 3. https://www.ncbi.nlm.nih.gov/pubmed/12042882.

Nakkach, Silvia. *Free Your Voice*. Louisville, CO: Sounds True, 2012.

Nakkach, Silvia. *Medicine Melodies: Song the Healers Hear*." Audio CD. Louisville, CO: Sounds True, 2011.

O'Connell, Cathal. "Topology explained—and why you're a donut." *Cosmos*. https://cosmosmagazine.com/physics/what-is-topology. Oct. 2016,

Ouspensky, P. D. *In Search of the Miraculous*. New York, NY: Harvest Books, 1949.

Pauli, Wolfgang. *Writings on Physics and Philosophy*. Edited by Charles P. Enz and Karl von Meyenn. New York, NY: Springer-Verlag, 1994.

Perls, Fritz. *Gestalt Therapy Verbatum*. Gouldsboro, ME: The Gestalt Journal Press, 1992.

Petsch, H. "Approaches to verbal, visual, and musical creativity by EEG coherence analysis." *International Journal of Psychophysiology 24*. 1996.

Ran, Weiyu, Steven Fredericks, and John R. Saylor. "Shape oscillation of a levitated drop in an acoustic field." Cornell University Library. https://arxiv.org/find/physics/1/au:+Saylor_J/0/1/0/all/0/1. 2013.

Reif, Andreas, Christian P. Jacob, Dan Rujescu, Sabine Herterich, Sebastian Lang, Lise Gutknecht, Christina G. Baehne, Alexander Strobel, Christine M. Freitag, Ina Giegling, Marcel Romanos, Annette Hartmann, Michael Rösler, Tobias J. Renner, Andreas J. Fallgatter, Wolfgang Retz, Ann-Christine Ehlis, and Klaus-Peter Lesch. "Influence of Functional Variant of Neuronal Nitric Oxide Synthase on Impulsive Behaviors in Humans." *Arch. Gen. Psychiatry Archives of General Psychiatry*. https://pdfs.semanticscholar.org/8d7f/7db1731987a4a271078f72d61ee9ecfd7deb.pdf. 2009.

Reimann, Michael W., Max Nolte, Martina Scolamiero, Katharine Turner, Rodrigo Perin, Giuseppe Chindemi, Pawel Diotko, Ran Levi, Kathryn Hess, and Henry Markram. "Cliques of Neurons Bound into Cavities Provide a Missing Link Between Structure and Function." *Frontiers in Computational Neuroscience*. June 2017.

Riganello, F., A. Candelieri, M. Quintieri, and G. Dolce. "Heart Rate Variability, Emotions, and Music." *Journal of Psychophysiology*. 2010.

Riley, Terry. *The Trinity of Eternal Music—Sound and Light: La Monte Young & Marian Zazeela*. Edited by William Duckworth & Richard Fleming. Cranbury, NJ: Associate University Presses, 1996.

Robins, Lee N. "Vietnam veterans' rapid recovery from heroin addiction." St. Louis, MO: Washington University School of Medicine. http://www.rkp.wustl.edu/veslit/robinsaddiction1993.pdf. 1993.

Roob, Alexander. *Alchemy & Mysticism*. New York, NY: Taschen, 1997.

Rosen, Jeffrey B. Donley, Melanie P. "Animal studies of amygdala function in fear and uncertainty: Relevance to human research." *Biological Psychology,* July 2006.

Rudhyar, Dane. *The Magic of Tone and the Art of Music*. Boulder, CO: Shambhala, 1982.

Rüütel, Eha. "The Psychophysiological Effects of Music and Vibroacoustic Stimulation." *Nordic Journal of Music Therapy*. http://www.tandfonline.com/doi/bs/10.1080/08098130209478039. 2002.

Schneider, M. *A Beginner's Guide to Constructing the Universe: The Mathematical Archetypes of Nature, Art, and Science*. New York, NY: Harper Collins, 1994.

Schwenk, Theodor. *Sensitive Chaos: The Creation of Flowing Forms in Water & Air*. New York, NY: Schocken Books, 1976.

Scott, Cyrill. *Music: Its Secret Influence Through the Ages*. London: Theosophical Publishing House, 1937.

Smith, William, and Samuel Cheetham. *A Dictionary of Christian Antiquities*. London: John Murray, 1875.

Stapp, Henry P. "Quantum approaches to consciousness." *The Cambridge Handbook of Consciousness*. New York, NY: Cambridge University Press, 2007.

Stefano, G. B., G. L. Fricchione, B. T. Slingsby, and H. Benson. "The Placebo Effect and the Relaxation Response: Neural Processes and their Coupling to Constitutive Nitric Oxide." *Brain Research Reviews* 35. 2001.

Stone, R. *Polarity Therapy: The Complete Works. Vol. II*. Reno, NV: CRCS Publications, 1987.

Strassman, Rick. *DMT: The Spirit Molecule*. Rochester, VT: Park Street Press, 2001.

Sutherland, W. G. *Contributions to Thought*. Fort Worth, TX: Rudra Press, 1967.

Tufail, Yusuf, Anna Yoshihiro, Sandipan Pati, Monica M Li, and William J. Tyler. "Ultrasonic neuromodulation by brain stimulation with transcranial ultrasound." *Nature Protocols*. 2011.

Tzu, Lao. *Tao Te Ching*. CA: Sacred Books of the East, 1891.

Upledger, J., and J. Vredevoogd. *Craniosacral Therapy: Vol. I*. Seattle, WA: Eastland Press, 1983.

Wass, Caroline, Daniel Klamer, Kim Fejgin, and Erik Palsson. "The Importance of Nitric Oxide in Social Dysfunction." *Behavioural Brain Research*. http://europepmc.org/abstract/med/19166879. 2009.

Weiming, Xu, Li Zhi Liu, Marilena Loizidou, Mohamed Ahmed, and Ian Charles. "The role of nitric oxide in cancer." *Cell Research*. 2002.

Wiese, B. S. "Successful pursuit of personal goals and subjective well-being." *Personal Project Pursuit: Goals, Action and Human Flourishing*. Editors: B. R. Little, K. Salmela-Aro, and S. D. Phillips. Mahwah, NJ: Lawrence Erlbaum, 2007.

Little, B. R., Katarina Salmela-Aro, and Susan D. Phillips, editors. *Personal Project Pursuit: Goals, Action, and Human Flourishing*." Mahwah, NJ: Lawrence Erlbaum. http://psycnet.apa.org/record/2006-11798-000. 2007.

Xenakis, Iannis. *Formalized Music: Thought and Mathematics in Composition*. Hillsdale, NY: Pendragon Press. 1992.

Acknowledgements

There are many people who have supported and helped me in many ways over many years to complete *Sound Healing and Values Visualization*.

First and foremost, I want to thank all of my students for being there and asking lots of questions. My students are my teachers.

My wife Thea Keats Beaulieu and my twin sons Daniel and Lucas, the intervals of my life.

My son Lars Beaulieu and my granddaughters Lua and Lucienne, the overtones of my life.

My Mother and Father for giving me everything I needed to be myself.

Pamela Kersage for her years of support and always being there.

Sathya Sai Baba for spiritual guidance.

Randolph Stone, D.O, D.C., N.D., for creating Polarity Therapy energy principles.

I want to thank all the musicians who have visited my Stone Ridge studio for our sound journeys that have enriched my life with great sounds and discussions.

I want to thank the musicians and poets of Axial Band, George Quasha, Charles Stein, and David Arner for our adventures into sound.

William R. Howell, Esq., for opening up a new vista where both science and energy can be integrated.

Peter and Julie Wetzler for inspiring everyone around them to be creative.

Bob and Mary Swanson (Riverbank Laboratories) who keep us tuned.

Gerry and Barbara Hand Clow for their friendship, vision, and energy counsel.

Philippe Garnier (Sonothérapeute, Paris), master of crystal.

Lea Garnier (SAGE Foundation, Woodstock, NY), guider of sound beings.

Gary Strauss and Tracy Griffiths (Polarity Healing Arts Los Angeles California) for their support and creating a healing center which represents the essence of Polarity Therapy.

Andreas and Brigitta Raimann Lederman (Schule fur Holistische Naturheilkunde Zug, Switzerland) for their many years of friendship and support.

Urs and Paki Honauer (Polarity Zentrum, Zurich, Switzerland) for their many years of friendship and support.

Jeff Volk (Macromedia) for his many years of support and sound discussions.

Jonathan and Andi Goldman (Healing Sounds) who hum mystical tuning forks.

Dave Stein (*Rose+Croix Journal*) for his amazing editing and patience.

Swami Srinivasan, Swami Swaroopananda, Swami Bramananda, Swami Sita, and the all the staff of the Sivananda Yoga Centers for their spiritual support, vision, and great Carribean conversations.

Silvia Nakkach (Vox Mundi Berkley, California) for sharing her love of music and sound and inspiring all around her.

David Perez-Martinez, M.D., who leads the way in sound healing and medicine.

Aruna Bakhru, M.D., who inspires and opens new doorways for sound healing and integrative medicine.

Mitch Nur (9Ways Academia) for his great stories and love of Sacred Sound.

Sara Auster (All Good Sounds) for her pioneering work in sound healing education.

Professor Jean Kopperud (Univerity of Buffalo Music Department) for her continuing support in all that is artistic and musical.

Michael Alicia (Center for Advanced Therapeutic Arts) for his support of sound healing in the massage community.

Jorge Alfano for music and more music and more music…

Ivy Ross for her creative spirit and opening my eyes to sound healing in corporate reality.

Audrey Cusson and Jeff Culule (Mirabi of Woodstock) for all their support of sound healing and all that is good.

Joseph Schmidlin, D.O., (Health Touch) for his work integrating osteopathy and sound sound healing.

Paul Campbell (Sound Healer) for many years of support and collaboration.

Zorka Grigoro and Daniel Kreier (Ecole de Polarity Suisse Romande Geneva Switzerland) for their continuing inspiration and support.

Joachim Topler (Klangschwingung) for his ever multiplying crystal vibrations.

Susanne Degendorfe (IAK: Institute of Applied Kinsiology, Germany).

Susan Imholz, Ph.D., for inspiring me with her vision of Design Psychology.

Phil Young and Morag Campbell (International Polarity Education Association) who continue to inspire and open doorways for energy medicine.

Leslie McGuirk who inspires creativity in the stars.

John Chitty who knows it don't mean a thing if you don't have that swing.

Thomas Workman who plays sounds that transcend time.

Steve Gorn for our magical flute journeys into sonic realities.

Karen Kelly, D.C., for bringing the spirit of Chiropractic to sound healing.

Gil and Ellen Goldstein for island music.

Garry Kvistad (Woodstock Chimes) where the beat goes on and on.

Mike and Galena Tamburo (Crown of Eternity) for their wonderful gong journeys.

Peter Blum (Sounds for Healing) for inspiring trances.

Suzanne Durst for always being there and opening multiple doorways.

Ann Marie Cushing for her vision of sound healing today.

Jack and Lydia DeJohnette for always being sound.

Vicki Genfan who sings tuning forks.

Burkhard Behm (KiOase) for giving me a tuned German voice.

Mitchel Gaynor, M.D., for his support and contributions to the field of Sound Healing.

Martha Flaherty who shows everyone the way with crystal clarity.

Niloofar Shaterian and Bray Ghiglia (Transformational Arts Technologies) who bring peace through the arts.

Wendy Young who organizes sound healing like a harpsichord.

Stephanie Rooker (Voice Journeys) who sings tuning forks.

Manda Stretch (Very Fairy Events) a rare fairy who flutters to magic sounds.